T0205863

Taking the "Oof!" Out of Proofs

This book introduces readers to the art of doing mathematical proofs. Proofs are the glue that holds mathematics together. They make connections between math concepts and show why things work the way they do. This book teaches the art of proofs using familiar high-school concepts, such as numbers, polynomials, functions, and trigonometry. It retells math as a story, where the next chapter follows from the previous one.

Readers will see how various mathematical concepts are tied and will see that mathematics is not a pile of formulas and facts; rather, it has an orderly and beautiful edifice.

The author begins with basic rules of logic and then progresses through the topics already familiar to the students: numbers, inequalities, functions, polynomials, exponents, and trigonometric functions. There are also beautiful proofs for conic sections, sequences, and Fibonacci numbers. Each chapter has exercises for the reader.

Reviewer Comments:
I find the book very impressive. The choice and sequence of topics is excellent, and it is wonderful to have all of these things together in one volume. Theorems are clearly stated, and proofs are accurate. – *Michael Comenetz*

The thoroughness of the narrative is one of the main strengths of the book. The book provides a perfect illustration of mathematical thinking. Each step of a given derivation is precise and clear. – *Julie Gershunskaya*

Draganov's book stands out from the many competing books. Draganov's goal is to show that mathematics depends on the notion of proof. Unlike other transition books, he addresses mathematical topics at an accessible level, rather than topics studied later in the university curriculum. – *Ken Rosen*

Alexandr Draganov holds a PhD in Electrical Engineering from Stanford. After a career in high-tech, he pivoted to teaching and writing.

Textbooks in Mathematics
Series editors:
Al Boggess, Kenneth H. Rosen

Applied Differential Equations
The Primary Course
Vladimir A. Dobrushkin

Introduction to Computational Mathematics: An Outline
William C. Bauldry

Mathematical Modeling the Life Sciences
Numerical Recipes in Python and MATLAB™
N. G. Cogan

Classical Analysis
An Approach through Problems
Hongwei Chen

Classical Vector Algebra
Vladimir Lepetic

Introduction to Number Theory
Mark Hunacek

Probability and Statistics for Engineering and the Sciences with Modeling using R
William P. Fox and Rodney X. Sturdivant

Computational Optimization: Success in Practice
Vladislav Bukshtynov

Computational Linear Algebra: with Applications and MATLAB® Computations
Robert E. White

Linear Algebra With Machine Learning and Data
Crista Arangala

Discrete Mathematics with Coding
Hugo D. Junghenn

Applied Mathematics for Scientists and Engineers
Youssef N. Raffoul

Graphs and Digraphs, Seventh Edition
Gary Chartrand, Heather Jordon, Vincent Vatter and Ping Zhang

An Introduction to Optimization with Applications in Data Analytics and Machine Learning
Jeffrey Paul Wheeler

Encounters with Chaos and Fractals, Third Edition
Denny Gulick and Jeff Ford

Differential Calculus in Several Variables
A Learning-by-Doing Approach
Marius Ghergu

Taking the "Oof!" Out of Proofs
Alexandr Draganov

https://www.routledge.com/Textbooks-in-Mathematics/book-series/CANDHTEXBOOMTH

Taking the "Oof!" Out of Proofs

Alexandr Draganov

CRC Press is an imprint of the
Taylor & Francis Group, an **informa** business
A CHAPMAN & HALL BOOK

First edition published 2024
by CRC Press
2385 Executive Center Drive, Suite 320, Boca Raton, FL 33431

and by CRC Press
4 Park Square, Milton Park, Abingdon, Oxon, OX14 4RN

CRC Press is an imprint of Taylor & Francis Group, LLC

ISBN: 978-1-032-59902-1 (hbk)
ISBN: 978-1-032-59598-6 (pbk)
ISBN: 978-1-003-45676-6 (ebk)

DOI: 10.1201/9781003456766

Typeset in CMR10
by KnowledgeWorks Global Ltd.

Publisher's note: This book has been prepared from camera-ready copy provided by the authors.

To Anna and Andrew

...we know well what is the
object of mathematics, and that
it consists in proofs...

Pascal

Let proof speak

Shakespeare

Contents

List of Figures

Preface

If you ask people on the street what math is about, many would say that it is a discipline about numbers. (Some would even add that they "are not good with numbers.") However, if you open a random paper in a math journal, you would hardly see any numbers. So, what *is* math really about and what differs math from any other discipline?

The etymology of the word *mathematics* goes back to a Greek word that means *to learn*. This suggests that mathematics deals with learning as such, and not with any particular subject. Indeed, from that paper in a math journal we would learn something new about some abstract mathematical objects (whether numerical or not). Of course, the same can be said about many papers in natural sciences or in humanities: we learn something new. One major difference is that in math any new knowledge is established using rigorous *proofs*. That's why proofs are an indispensable part of mathematics.

This book will introduce you to proofs in mathematics by loosely following the material from the high-school curriculum. Students who decide to major in math in college discover that all math classes there contain a serious dose of proofs. Unfortunately, this often comes as a surprise and a challenge. Standard high-school math textbooks pay lip service to proving formulas or other mathematical facts: there may be some proofs in geometry, and a textbook on calculus may present a few isolated proofs. The goal of most high-school textbooks, however, is to give students basic practical skills for solving equations, computing derivatives, and such. To apply a mathematical formula to a real-world problem, it may not be necessary to know where that formula comes from or how it can be proved, as long as we trust it. As a result, students learn a scattered collection of recipes to solve problems, such as the quadratic formula and trigonometric identities, but the proofs that hold all that knowledge together remain hidden. In this book, we will use these familiar concepts (including the quadratic formula and trigonometric identities) to learn about how proofs are made.

You may then ask: why do we need to learn about proofs if some smart mathematicians have already proved stuff for us and if we can get practical quantitative skills without ever seeing that proof thing? I have four answers to this question.

Two of these answers deal with applying math and can be explained using an analogy with flying a plane.

1. I once spoke with a professional pilot and he described to me the true challenges of his job. He said that as someone learns to fly a plane, they start during daylight and in good weather. At night or in the fog, the job of flying a plane requires advanced skills. In particular, the pilot must not trust their subjective feeling of pitch, yaw, and roll (the angles the plane makes with the horizon). Instead, the pilot must fully defer to the instruments. Because of the dynamics of the flight, the pilot's own sense of orientation in space becomes unreliable.

   When exploring new phenomena or analyzing data, we are not unlike a pilot who flies a plane in the dark. Sometimes we must suspend our intuition and just go where the logic takes us. This book will have many examples of counter-intuitive, yet true statements. Relying on intuition too much makes you prone to mistakes. When every step is proved, you can be confident in your answer: you will know that your conclusion is true and correct.

2. Another lesson from that conversation with the pilot was that modern passenger jets are highly automated, and this makes learning the basic takeoff, cruising, and landing fairly easy. The real challenge is to learn how to handle special, unforeseen situations (including emergencies).

   The same can be said about using math to solve practical problems. In many cases, the set of skills that a student gets in standard math classes is sufficient to perform common tasks in engineering, economics, or data analysis. In fact, many such skills can be automated, like using the autopilot in a jet plane. For example, the computation of derivatives no longer has to be done with pen and paper, but one can simply plug a function into a tool like Wolfram Alpha. The value of human thinking comes from the ability to handle special, nonstandard cases. Too many students who took calculus in high school and learned there about that wonderful Taylor series do not know that it does not work for many functions, even if these functions are infinitely differentiable. Luckily, mathematical proofs always delineate the domain of applicability for each statement. A careful observation of a proof shows where a formula works and where it does not.

Two more reasons to learn about mathematical proofs go beyond the practical applications of mathematics.

3. Knowing how to flip on and off the logic switch in your brain is a good skill to have. That way, you can see a hidden contradiction in a politician's speech, or an unfounded claim in a marketing brochure. Logic is a great (but not the only) means to guard yourself from being taken advantage of. Proofs teach you to apply logic in a rigorous, systematic way.

4. And, last but not least: many proofs are just too beautiful not to study. They show you a harmony in the world you live in, uncover unexpected connections, or create the feeling of a breakthrough. Beauty pops up when parts miraculously fall into place, creating a coherent, intelligible picture. Going through a good proof is like climbing a mountain: you begin by putting in a lot of effort, and then there is a moment when you step onto the top and see a great panorama laid out in front of you. Suddenly, everything makes sense, as if the jigsaw puzzle of farming plots and people's hamlets at the foot of the mountain is now complete. At that moment you tell yourself: it's been worth the climb!

   The French writer Marcel Proust once wrote in passing, as if saying something obvious, that beauty follows from orderly complexity. That may be the source of beauty in mathematics: it shows the order in complexity.

This book will show you how to do proofs by going through more than 100 theorems. I tried to select the topics for the book in such a way that they relate to each other. Some theorems serve as a foundation for proving other theorems, with all of them steadily telling a coherent story. The material does not venture beyond the level of a high-school algebra and trigonometry curriculum. Together, we will go through properties of numbers, inequalities, functions, polynomials, and other familiar topics. I hope that after reading this book you will be able to say: "now that formula from the high-school algebra class finally makes sense."

My own introduction to proofs in mathematics started in high school, when the most talented teacher, Ms. Vera Rosenberg, showed us the beauty of rigorous mathematics, including the stunning Georg Cantor's "diagonal proof" (it is presented in section 2.8.3 of this book). I am also grateful to other great teachers, Drs. Yuri Gorgo, Nickolay Kotsarenko, Umran Inan, and Tim Bell. Professor Michael Comenetz made many helpful comments on the manuscript. Two nonstandard and curious proofs that I learned from Professor Michael Comenetz are included in this book (a proof that the number of primes is infinite on p. 23 and a proof of irrationality of $\sqrt{2}$ on p. 37). Writing this book would not be possible without the patient support and help from my wife, Luda, our children, Anna and Andrew, and our son-in-law, Daniel Palmer. The work of my late father, Boris Draganov, and the books published by him were my inspiration for this project.

1

A Few Rules of Logic

1.1 True and False Statements

A proof establishes that a particular mathematical statement is true. To learn about proofs, we first need to understand how to operate with true and false statements, and this chapter is a primer for that.

Any statement can be either true or false.[1] There are no "partially true" statements. For example, we may solve equation $5x = 0$ to determine that $x = 0$. Then the statement $x = 0$ is true. We can denote this statement as A and write $A = T$, where T denotes a true statement.

For any statement A there is a negation, "not A," which we denote $\neg A$. In the above example, the negation of statement $x = 0$ would state that $x \neq 0$. If a statement is true, its negation is false, and the other way around. We also state that $\neg(\neg A) = A$.

Multiple statements can be combined using "or" and "and" operators. For example, we may solve equation $x^2 - 5x + 6 = 0$ to find two roots. Then, we can say that $x = 2$ *or* $x = 3$. If we denote the first statement $x = 2$ as A and the second statement $x = 3$ as B, the combined statement will be written as $A \vee B$, where the symbol $\vee$ denotes logical "or." In the above example, x cannot be equal to both 2 and 3, but in general statement, $P \vee Q$ means that P is true, Q is true, or both.

We can also say that in the above equation $x < 5$ *and* $x > 0$. If these two statements are denoted as C and D, then we can write the combined statement as $C \wedge D$, where the symbol $\wedge$ denotes logical "and."

If we know that a part of a combined statement is true or false, we can either make a conclusion about the whole combined statement, or simplify it. If we think about it, rules for doing that are intuitive. For any statement A, we conclude that:

$$\begin{aligned} T \vee A &= T, \\ F \vee A &= A, \\ T \wedge A &= A, \\ F \wedge A &= F. \end{aligned} \tag{1.1}$$

[1] To be exact, there are statements, for which it is impossible to determine if they are true or false, but we will not deal with those.

DOI: 10.1201/9781003456766-1

Indeed, it is enough for one statement in the pair to be true for a combined "or" statement to be true. If one statement in the pair is false, the combined "or" statement completely hinges on the second statement in the pair. For combined "and" statements, one true statement in the pair is redundant, and one false statement makes the whole "and" combination false.

1.2 General and Particular Cases

Often statements apply to multiple cases or to some set of numbers; we will call such statements *general*. An example of such a statement would be "all integers that are multiples of 10 are divisible by 5." There are infinitely many numbers that are covered by this statement. For example, number 270 is a multiple of 10, which means that it is divisible by 5. This last statement about the number 270 is a particular case that follows from the general statement about all integers that are divisible by 10.

We can illustrate this example graphically. Figure 1.1 shows a universe of integers. Some of them are divisible by 5. Some are multiples of 10. Note that if a number is multiple by 10, it is always divisible by 5. However, not all numbers that are divisible by 5 are multiples of 10. Therefore, the statement "all integers that are divisible by 5 are multiples of 10" is false.

Let's introduce notations for the last two statements. The statement "all integers that are multiples of 10 are divisible by 5" we denote as P; this statement is true. The statement "all integers that are divisible by 5 are multiples of 10" we denote as Q; this statement is false. We can now formulate several important lessons:

1. If a general statement is true, then any particular case is true. For example, from statement P and from the fact that 270 is a multiple of 10, we can conclude that 270 is divisible by 5.

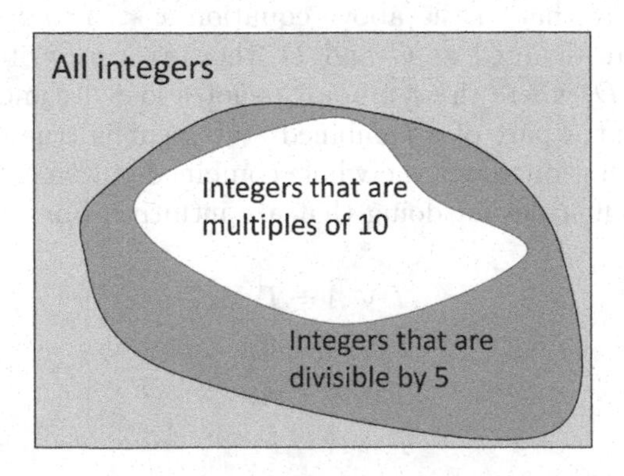

FIGURE 1.1
A graphic representation for a logic statement

2. Just one counterexample is enough to make a general statement false. For statement Q, we note that number 5 is divisible by 5 but is not a multiple of 10. Therefore, statement Q is false. In mathematics, we do not say that a statement is "sometimes false" or "partially false." If we have just one counterexample, the statement is false, period.

A general algebraic formula may apply to all values of variables or be valid only for some values. For example, $2x + b = 0$ is true only if $x = -b/2$, but $(2x + b)^2 = 4x^2 + 4xb + b^2$ is valid for all values of x and b. If an algebraic formula is true for all values of variables, it is called an identity.

1.3 "If... then" Statements

Statement P that were introduced in the previous section can be written in a different form: "if an integer is a multiple of 10, then it is divisible by 5." Such "if ... then ..." statements are very common in proofs. Let's denote the first half ("an integer is a multiple of 10") as K, and the second half ("it is divisible by 5") as L. Then the whole statement P will be written as $K \implies L$, where the symbol $\implies$ reads as "implies that." The implication has the transitivity property: if $K \implies L$ and $L \implies M$, then $K \implies M$. Statements like $K \implies L$ are called *conditional* because K is a condition for L.

Statement Q can be written as $L \implies K$. Note that statements K and L are swapped here; we say that statement Q is *converse* to statement P. It is important to remember that if for any pair of statements K and L the statement $K \implies L$ is true, the converse statement $L \implies K$ is not necessarily true. (We just saw a case when the converse statement is false.)

Let's look at the negation for statements K and L. For the former, the negated statement $\neg K$ would be "an integer that is *not* a multiple of 10." In figure 1.1, these would be all points that lie outside of the white area containing all integers that are multiples of 10.

For the latter, the negated statement $\neg L$ would be "that integer is *not* divisible by 5." In the figure, these would be all integers that lie outside of the darker area that contains all integers that are divisible by 5.

We can see that some points outside of the white area (where K is false) fall into the darker area (where L is true). Therefore, statement $L \implies K$ is false. (Again, it is not partially false, it is false, period!)

The situation changes if we consider a statement that is called *contrapositive*. We know that our original statement $K \implies L$ is true. The contrapositive statement applies negation to both parts and swaps them: $\neg L \implies \neg K$. This statement will be as follows: "if an integer is not divisible by 5, then it is not a multiple of 10." If we think a minute about this statement, we will

TABLE 1.1
Types of conditional statements.

Statement	Relation to statement $K \implies L$	Equivalent to
$K \implies L$	self	$\neg L \implies \neg K$
$L \implies K$	converse	$\neg K \implies \neg L$
$\neg K \implies \neg L$	inverse	$L \implies K$
$\neg L \implies \neg K$	contrapositive	$K \implies L$

see that it must be true. In the figure, the "if" clause applies to the points that lie outside of the darker area (integers that are not divisible by 5). The second clause applies to the numbers that lie outside of the white area. This includes some numbers in the darker area and some outside of it. Therefore, any number that lies outside of the darker area also lies outside of the white area, making the contrapositive statement true. We conclude that if $K \implies L$ is true, then $\neg L \implies \neg K$ is also true. This can be written as follows:

$$(K \implies L) \implies (\neg L \implies \neg K). \tag{1.2}$$

However, $L \implies K$ is false for this particular example and may or may not be true for other examples:

$$(K \implies L) \not\Rightarrow (L \implies K). \tag{1.3}$$

Finally, for $K \implies L$ there is an *inverse statement*: $\neg K \implies \neg L$. By analogy to equation (1.2), we conclude that it is equivalent to $L \implies K$. All four types of statements are listed in table 1.1.

1.4 Combining "or," "and," and Negation

Let's now illustrate logical "and" and "or" statements (figures 1.2 and 1.3). We start from two statements: statement M says that number x is a multiple of 2, and statement N says that number x is a multiple of 5. Figure 1.2 shows that if x is a multiple of 2 *and* 5 (which implies that it is a multiple of 10), it must be located in the overlap of the two curves—that is, in the white area.

Next, consider some other number y, such that it is a multiple of 2 *or* a multiple of 5. (Note that one option does not exclude another: y still can be a multiple of both 2 and 5.) Figure 1.3 shows that number y must be located somewhere within one or both two closed curves.

What happens if we apply negation to statements $M \wedge N$ or $M \vee N$? In the figure, a negated statement corresponds to the area that is complementary to

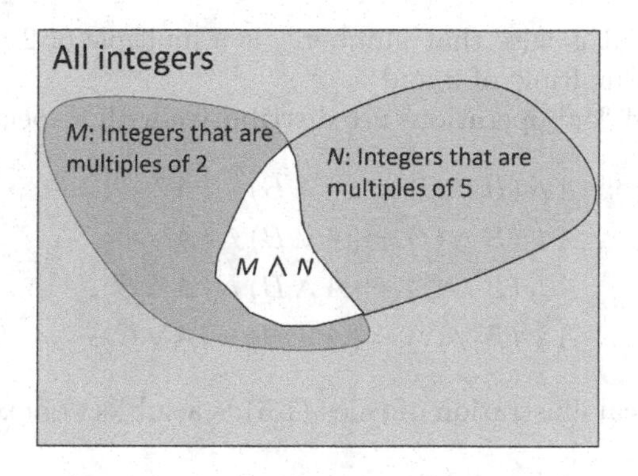

FIGURE 1.2
A graphic representation for a logical "and"

that for the original statement (lies outside of the area defining the original statement). In figure 1.2, this is the nonwhite area. It is given by

$$\neg(M \wedge N) = \neg M \vee \neg N, \tag{1.4}$$

For figure 1.3, the negation of statement $M \vee N$ would correspond to the area outside of the two closed curves. It is given by

$$\neg(M \vee N) = \neg M \wedge \neg N. \tag{1.5}$$

We see that the negation of an "and" or "or" statement does two things: it negates statements M and N and replaces "and" by "or" or "or" by "and."

In the above example, if it is false that number x is a multiple of 2 and a multiple of 5, then this number is not a multiple of 2 or not a multiple of 5

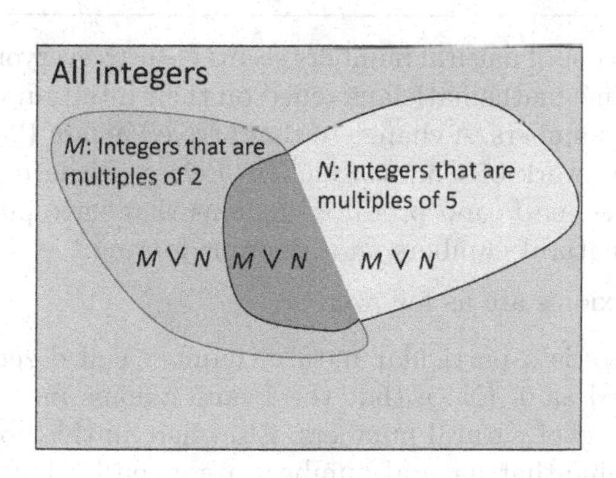

FIGURE 1.3
A graphic representation for a logical "or"

(or both). Also, if it is false that number y is a multiple of 2 or a multiple of 5, then it is not a multiple of 2 and 5.

The "and" and "or" operations are distributive with respect to each other:

$$\begin{aligned} A \wedge (B \vee C) &= (A \wedge B) \vee (A \wedge C), \\ A \vee (B \wedge C) &= (A \vee B) \wedge (A \vee C), \\ A \wedge (B \wedge C) &= (A \wedge B) \wedge (A \wedge C), \\ A \vee (B \vee C) &= (A \vee B) \vee (A \vee C). \end{aligned} \tag{1.6}$$

Building a graphical illustration of rules (1.6) is a subject of exercise 1 for this chapter.

1.5 Logic Lingo

We are already familiar with several terms from logic that we will be using in proofs. Here, we introduce a few more (in the alphabetical order):

1. *Axiom.* A typical theorem (a statement to prove) is based on one or more previous theorems. Of course, the succession of theorems referring to one another cannot be infinite and must have an origin. This means that there must be some statements that are accepted without proof. These statements are called *axioms* and are assumed to be true. By starting from a different set of axioms—that is, from different starting assumptions, one can build an entirely new math (and people have done so), like a new building can be constructed on a separate foundation.

The concept of natural numbers seems clear to everyone, and for a long time mathematicians relied on their intuition when dealing with numbers. A change to that arrived in late 19th century, when the work of Charles Sanders Peirce, Richard Dedekind, and Giuseppe Peano produced axioms that have put the concept of natural numbers on a rigorous footing.

Peano axioms are as follows:

(a) There is a particular natural number called zero and denoted as 0. (Note that the Peano axioms include zero in the set of natural numbers. Elsewhere in this book we will assume that natural numbers are given by $1, 2, 3, \ldots$, and zero is not a part of this set.)

(b) There is a *successor operation*: for any natural number n, there exists another number $S(n)$. In plain English, 1 is a successor of 0, number 2 is a successor of 1, and so on.

(c) However, zero itself is not a successor of any natural number.

(d) If two numbers are different, their successors are also different. There is no natural number that is produced as a successor of two different numbers.

(e) Finally, if we take zero and continue applying the successor operation, that would produce *all* the natural numbers—that is, any natural number can be produced by applying the successor operation multiple times.

Our goal, however, is not to build a meticulous apparatus of arithmetic from axioms, but just to show how proofs work. That is why we will skip many arcane steps and try to stay close to the topics that are familiar and applicable to many readers: algebra, functions, trigonometry, and so on.

2. *By definition.* This is a common way to say that a particular statement directly follows from the definition. For example, a proof may include the following argument as an intermediate step: "Let n be a natural number. By definition, n is positive." Here, we use the fact that all natural numbers $(1, 2, 3, \ldots)$ are positive.

3. *Conjecture* is a statement that is offered as tentatively true but is not yet proved. A conjecture presents a challenge to mathematicians, and many important breakthroughs in mathematics were achieved by trying to prove (or disprove) various conjectures. Often, there are strong arguments in favor of a conjecture that still fall short of a definitive proof. Even though mathematicians may deem a particular conjecture to be "likely true," that does not satisfy the high standards of rigor.

Conjectures are important for mathematicians. In 1859 the German mathematician Bernhard Riemann formulated a conjecture about nulls of the so-called zeta function. This conjecture is known as the Riemann hypothesis. Decades later, another great German mathematician, Davis Hilbert, said: "If I were to awaken after having slept for a thousand years, my first question would be: Has the Riemann hypothesis been proven?"

The Riemann hypothesis remains unproven today.

4. *Conjunction* is a statement that is the result of applying the "and" operator to two or more statements. For example, statement A says that $x < 0$, and statement B says that $x > -1$. The conjunction of these two statements is $A \wedge B$, which can be expressed as $-1 < x < 0$.
5. *Contradiction* occurs when a statement is deemed both true and false, such as when we state that both $2+2=4$ and $2+2 \neq 4$ hold. Contradictions are not allowed in mathematics (see section 1.6) but are widely used as a tool for proving theorems (see section 1.9)!
6. *Corollary* is a statement that follows from a particular theorem via a short proof. Since a corollary is proved, it can be viewed as a special kind of theorem. The distinction between a genuine theorem and a corollary is subjective and is based on how tight the connection is with the theorem, of which the statement in question is a corollary. For example, in section 6.1 we will prove that for any real x and natural n, m, it is the case that $(x^n)^m = x^{nm}$. From this theorem, we can formulate a corollary: for any real x and natural n, m it is the case that $(x^n)^m = (x^m)^n$. (Indeed, according to the original theorem, both sides of the last equation have the same value x^{nm}.)
7. *Definition.* A proof is impossible without defining *what* exactly we are proving. All mathematical objects must be precisely defined, and sloppiness in definitions leads to mistakes or – oh, the horror! – contradictions. In this book, we gloss over many definitions; for example, we do not define what natural numbers are. We do this because we are just peeking behind the curtain of mathematics rather than trying to get a full and comprehensive picture of it.
8. *Disjunction* is a statement that is the result of applying the "or" operator to two or more statements. For example, statement C says that $x < 0$, and statement D says that $x = 0$. The disjunction of these two statements is $C \vee D$, which can be expressed as $x \leq 0$.
9. *"For all ..."* or *"For each ...."* This phrase precedes a logical statement and means that a statement is true for all elements from a certain set.[2] For example, we can say that "for all values of b, equation $5x = b$ has only one solution, which is given by $x = b/5$." A phrase like "for all" is called a *quantifier* because it tells us how many individual elements from a given set satisfy a particular logical statement (in this case, all of them). For this quantifier, we use notation $\forall$, which is just the first letter of the word "All" turned upside down.

   Let's denote statement $5x = b$ as E, and statement $x = b/5$ as S. Then, we can write $\forall b(E \implies S)$—that is, "for all values of b, it follows from $5x = b$ that $x = b/5$."

[2] We will talk more about the terms "element" and "set" in section 2.8.1. In this chapter, we rely on an intuitive understanding of these terms.

Note that here we assume that x and b are numbers. In fact, a quantifier like $\forall$ always applies to elements of a particular set, whether implicitly or explicitly. For example, we can rewrite this quantifier as $\forall x, b \in \mathbb{R}$, where symbol $\in$ states that variables x and b belong to set $\mathbb{R}$, and that last symbol denotes the set of all real numbers.

Negation of a statement that is true for all objects in a set would mean that this condition is not true for at least one of the elements. For example, we can state that "all cats are grey." A negated statement will assert that there is (that is, there exists) at least one cat that is not grey.

10. *"If and only if ...."* We have seen how the " if P, then Q" construction works. We also noted that if $P \implies Q$ is true, that does not always mean that $Q \implies P$ is true. The expression "P if and only if Q" means that both $P \implies Q$ and $Q \implies P$ are true. For example, an integer is a multiple of 10 if and only if the last digit in its decimal representation is zero. Sometimes we may see an alternative term: instead of saying "P if and only if Q" mathematicians would say "P iff Q." Such statements are called *biconditional* because P is a condition for Q and Q is a condition for P.
11. *Implication* is such relationship between two statements, where one statement logically follows from another. For example, consider a conjunction of two statements from which it follows another statement. Let statement F say that $a \neq 0$ and statement G say that $ax = 0$. A conjunction of these two statements is $F \wedge G$. But, we know that if a product of two numbers is zero, at least one of them is zero, so we can conclude that $x = 0$. Let's denote this last statement as H. We say that $F \wedge G$ *implies* H. Another way of saying this is "if $(F \wedge G)$, then H." This statement is written as $(F \wedge G) \implies H$.
12. *Lemma* is an intermediate result, which is used as a step stone for proving a theorem. We may think of a lemma as a kind of a theorem, whose only application is to be used in proving another theorem.
13. *Necessary and sufficient conditions.* Sometimes we say that condition A is *necessary* for statement B to be true. This is a way of saying that if A is false (that is, if condition A is not satisfied), then B is false. This can be written as $\neg A \implies \neg B$, which is equivalent to $B \implies A$. For example, consider the following statement: "an integer being divisible by 2 is a necessary condition for the last digit of that integer to be equal to 6." This is the same as saying that if the last digit of an integer is equal to 6, that integer is divisible by 2.

    If condition C is *sufficient* for statement D to be true, this means that $C \implies D$. For example, "the last digit of an integer being zero is sufficient for that integer to be divisible by 5."

We may have a case when condition F is necessary and sufficient for statement G. That means that these two statements are equivalent. For example, "all angles of a triangle being equal is a necessary and a sufficient condition for this triangle to be equilateral." This means that all equilateral triangles have equal angles and that all triangles that have equal angles are equilateral. Such statements are called *biconditional.* It is the same as saying "F iff G."

When considering the statement of a theorem, it is important to carefully parse that statement and to determine each condition that it includes. Some of these conditions may not be immediately clear, but still crucial.

14. *Paradox* is a statement that runs contrary to our expectations or is apparently (but not always genuinely) contradictory. Unlike a bona fide contradiction, a paradox can be resolved. A paradoxical statement can be counter-intuitive but true (see Hilbert's Grand Hotel paradox on p. 54), or there may be a hidden flaw in the reasoning that has led to the paradox (for example, see exercise 13 for chapter 5). Interesting paradoxes occur when we deal with a statement that applies to itself, like the attempt to apply the barber's rule to himself in Russell's paradox (p. 12). Paradoxes, whether in math or in everyday life, can spur critical thinking.

15. *Theorem* is the statement that we are proving. Any proof must rely on some combination of the following: the definitions of the objects that the theorem pertains to, axioms, and one or more previously proved theorems. Tying a proof to something that we already know makes the whole structure of mathematics rigorous.

16. *"There exists a ..."* is another quantifier. The symbol for this quantifier is $\exists$, which is a flipped upper-case first letter of the word "Exists." This quantifier also requires us to specify a set, which that particular object is an element of.

    It is always important to establish the existence of a mathematical object. For example, we can say that "for all values of real numbers b, there *exists* such a real-valued x that $5x = b$." This can be written as $\forall b \in \mathbb{R}\ \exists x \in \mathbb{R}(5x = b)$. However, a similar statement would not be true for equation $bx = 5$ because its solution does not exist if $b = 0$.

    Quantifier $\exists x$ tells us that there exists at least one object x, but there can be two, three, or more such objects. We may still write $\forall c > 0 \exists x \in \mathbb{R}(x^2 + x - c = 0)$ because for any $c > 0$ equation $x^2 + x - c = 0$ has two real-valued solutions.[3]

[3]We will learn about a criterion for a quadratic equation to have real-valued solutions in section 5.6.

To specify that there exists exactly one object x, we use a different symbol: $\exists!x$. For example, we can state that $\forall b \in \mathbb{R}\exists!x \in \mathbb{R}(5x = b)$ because there is only one solution for equation $5x = b$ for any value of b. As a counterexample, it would be a mistake to use the symbol $\exists!$ to refer to the solutions of a quadratic equation.

A negation of a statement of existence would mean that such an object does not exist. Consider a statement, "there is at least one cat that can fly." Negation of this statement will assert that "there is no cat that can fly."

17. *"Without loss of generality ..."* This expression is used when we restrict the proof to a particular case, but still can map out a way to extend it to the general case.

    For example, consider a proof of the following statement: for any two numbers a and b such that $a \neq b$, the arithmetic mean $(a+b)/2$ is greater than one of these numbers and smaller than the other one. A proof of this statement could start as follows: "Without loss of generality, let us assume that $a < b$. Then ..." Though the original statement does not limit values of a and b to this inequality, using it can make the proof easier. After we have proved the desired statement for the case $a < b$, it is clear that we can do it in exactly the same way for the case $b < a$. Therefore, our constraint $a < b$ is just a temporary sacrifice, easing the path for the proof.

Now that we now what axioms and theorems are, we can revisit the notion of true and false statements. In mathematics, it is different from that in everyday life. A statement is "true" if it can be unambiguously proved using other previously proved statements, axioms, or both. A true statement for one set of axioms may turn out to be false for another set of axioms. This way of thinking about truth can be illustrated by an analogy. In Bulgaria, the gesture for "yes" is shaking one's head from side to side, which in the United States or France would mean "no." Within the confines of Bulgarian culture, people understand this gesture perfectly well, as they do it within the confines of American culture, albeit differently. Similarly, a statement can be deemed true or false only within the confines of a set of axioms.

1.6 No Contradictions Are Allowed

Mathematics and logic do not allow contradictions. No statement can be both true and false at the same time. In addition to being intuitive, this requirement is necessary because its violation would have immense consequences. Let us consider what would happen if we allowed just one statement Z to be both true and false. For example, consider what the world would look like if we allowed both $2+2=4$ and $2+2 \neq 4$.

Let's pair our contradictory statement Z with some other statement Y. So far we do not specify that other statement, nor do we know if it is true or false. However, since Z is true (in our example, $2+2=4$ is true), we know that the disjunction statement (Z or Y) is also true:

$$Z \vee Y = T. \tag{1.7}$$

Consider the negation of that last statement, which must be false:

$$\neg(Z \vee Y) = F. \tag{1.8}$$

We apply the rule from section 1.4 to get

$$\neg Z \wedge \neg Y = F. \tag{1.9}$$

However, we also know that $\neg Z$ is true (in our example, we also stated that $2+2 \neq 4$, due to the contradiction that we have allowed):

$$T \wedge \neg Y = F. \tag{1.10}$$

Next, we turn to equation (1.1) and conclude that

$$\neg Y = F, \tag{1.11}$$

which in turn means that $Y = T$. Now, we proved that statement Y is true.

The bizarre nature of this proof is that we have not specified *anything at all* about statement Y. Apparently, it can be *any* statement, for example, "a purple dinosaur has walked into the White House on November 9, 1961." This statement is proven to be true if we allow $2+2=4$ and $2+2 \neq 4$. Moreover, the statement "a purple dinosaur has *not* walked into the White House on November 9, 1961" also becomes true, making it contradictory too. Therefore, just one contradiction makes all statements contradictory. All our thoughts, arguments, and facts become useless. This shows why math and logic do not allow contradictions.

> The development of mathematics has not been perfectly smooth. On multiple occasions mathematicians ran into paradoxes or contradictions, which then had to be removed. Set theory (which we will briefly explore in section 2.8.1) originally featured a serious problem that can be illustrated using the famous Barber paradox, which is an example of a general paradox by Bertrand Russell.
>
> Consider a small town with one barber, who shaves all men, and only those men, who do not shave themselves. The question is, does the barber shave himself?

Suppose, he does. Since he shaves himself, he belongs to the group of men, whom he does not shave, which is a contradiction.

Now let's suppose he does not shave himself. Then he belongs to the group of men, whom he shaves, which is also a contradiction.

For this particular paradox we must simply conclude that such a barber does not exist. For the related set-theory problem (which this paradox is an illustration of) we must modify the axioms in such a way that this problem would not arise.

1.7 The Need for Existence

Many rules of logic were formulated by Aristotle (384-322 BCE). Scientists, philosophers, and mathematicians used Aristotelian logic for more than two millennia. In the second half of the 19th century, Gottlob Frege hit a reset button on the discipline of logic, particularly in its application to mathematics. Frege's ideas were further developed by Bertrand Russell. An important part of these new developments was a formal mechanism, which prevents errors in a naïve application of logic.

Consider the famous Russel's example: "The present king of France is bald." Since France is a republic, there is no present king of France, and this statement is false. Therefore, the negation of this statement must be true. A naïve application of negation produces the statement "The present king of France is not bald." However, this statement is false as well. This violates the fundamental principle of logic that if a statement is false, its negation must be true.

In Frege's and Russell's analysis, the original statement is parsed and analyzed in such a way, that the negation would state that there is no such person, who is the current king of France and bald.

Note that the difficulty with the naïve analysis of the statement "The present king of France is bald" arises from the fact that there is no king of France. In English, difficulties like this one are compounded by the ambiguity of the word *is*, which can mean existence (as in: "I think, therefore I am" by Descartes) or a predicate saying something *about* the subject (as in: "101 is a number"). We must carefully establish the existence (or lack of it) of every object we are dealing with. That's why there are so many proofs that simply establish the existence of a mathematical object, sometimes even without specifying anything concrete about that object. This and other types of proofs are listed in the next section.

1.8 What Is Typically Proved?

Proofs produce many different types of outcomes, but the most frequent ones are as follows:

1. *Existence (or lack of it).* This is self-explanatory: a proof shows that something exists (or it does not). For example, given an equation we prove that it has a solution. In a less-than-rigorous application of mathematics people sometimes fail to check for existence of the stuff they are trying to deal with, which often leads to incorrect or absurd results.
2. *Uniqueness (or lack of it).* If we have established that an object exists, the next question is whether it is unique. Does that equation have only one solution? Maybe there are two solutions? Or infinitely many?
3. *Formula or algorithm.* We prove that a particular way of doing something leads to a desired result. For example, we can prove the quadratic formula for solving quadratic equations.
4. *Property.* We prove that some mathematical object has a particular property. For example, in section 6.1, we will prove that exponentiation has the property $a^n a^m = a^{m+n}$ for any real a and any positive integers m and n.

1.9 Types of Proofs

Many proofs fall within a few categories that sometimes have special names. Here are the most common types of proofs:

1. *Considering all cases.* Sometimes the problem can be split into several cases, which are proved separately. For example, consider the following statement: "equation $(x-a)(x-b)=0$ has two solutions that are given by $x=a$ and $x=b$." If we proceed to solving this equation, we notice that the left-hand side is a product of two terms. In order for a product to be equal to zero, one or both of these terms must be zero. This splits the equation into two cases that can be considered separately: $x-a=0$ and $x-b=0$.
2. *Constructive proofs.* For an existence proof, we often point to or construct the object, whose existence we are proving. Same works for proving a formula if we can derive it.

A constructive proof is not the only way to establish existence. Some proofs do not pinpoint the object in question, but still prove its existence. We will see beautiful examples of such proofs in theorems 2.15 and 9.13.

3. *By contradiction.* To prove that some statement A is true, it is sufficient to prove that statement $\neg A$ is false. A proof by contradiction first deliberately assumes that statement $\neg A$ is true (even though we are trying to prove the opposite). Then, we build a sequence of logical arguments that results in a contradiction. Since contradictions are forbidden, we must conclude that our original assumption was wrong, and $\neg A$ is false. This proves statement A.
4. *By contrapositive.* From section 1.3 we know that statement $K \implies L$ and its contrapositive $\neg L \implies \neg K$ are equivalent. Therefore, to prove an implication statement we just need to prove its contrapositive.
5. *By induction.* This method is commonly used to prove statements that apply to multiple cases that can be enumerated. One example would be to prove a statement that is valid for all natural numbers (1, 2, 3, ...). For such a statement, we cannot consider all cases individually; instead, we do the following.

   To prove a statement that applies to all natural numbers, we first check if it is true for the first element—that is, for 1. Then, we prove that if the statement is true for some arbitrary number k, then it is true for the next number $(k + 1)$. By virtue of these two results, we can say that since the desired statement is true for $k = 1$, then it must be true for $k = 2$; next, it must be true for $k = 3$, and so on. The fifth Peano axiom (see p. 6) shows that this process will prove the statement in question for all natural numbers. Indeed, we start from case 1, then prove the statement for case 2, which is the successor of 1, and continue to move from one case to the next by applying the successor operation. By the Peano axioms, this will cover all natural numbers without exception.

   Sometimes we use the so-called *strong form* of mathematical induction. In this case we again start from proving a statement for $k = 1$, and then prove that if it is true for all $m \leq k$, then it is true for $(k + 1)$. The name of this method (strong induction) does not mean that it is a stronger or a more valid method than the basic form because both forms of this method are equally rigorous and powerful.

   Mathematical induction cannot be used for statements that apply to all real numbers because, as we will learn below, real numbers cannot be enumerated.

A proof does much more than simply establish a statement. It also connects the statement in question to other statements and concepts, helping shape the structure of mathematics as a whole. That is why it is often helpful to have multiple proofs of the same theorem done from different perspectives and even drawn from different areas. Paul Erdős, one of the most celebrated mathematicians of the 20th century, by his own admission knew 37 different proofs of the Pythagoras theorem at the age of 17. In this book, you will see several examples of when a theorem is proved in two or three different ways (but not in 37 ways, of course).

This chapter laid out a few rules of logic. They will be put to good use in the following chapters, where we will go through about 150 proofs in different topics, ranging from numbers, to trigonometry, to conic sections.

Exercises

1. Illustrate equations (1.6) graphically.
2. Show that in equations (1.6) the second equation is equivalent to the first, and the fourth equation is equivalent to the third.
3. Illustrate graphically the transitivity property of implication: if $A \implies B$ and $B \implies C$, then $A \implies C$.
4. Consider statements $A \implies B$ and $C \implies D$. Show that if C is a negation of A, then D implies a negation of B, and B implies a negation of D.
5. Prove that for any two statements A and B it is the case that $A = (A \wedge B) \vee (A \wedge \neg B)$.
6. Statement $A \vee \neg A$ is always true. Negate this statement and interpret the result.
7. Consider different cases to prove that for any two statements A and B we have $(A \vee B) \wedge \neg B = A \wedge \neg B$.
8. Consider different cases to prove that for any two statements A and B we have $(\neg A \wedge \neg B) \vee A = A \vee \neg B$.
9. Prove that $(A \vee B) \wedge B = (A \wedge B) \vee B = B$ for any A, B.

Exercises 10 through 15 use the following properties of inequalities: for any two real numbers c, d, statement $c > d$ is a negation of the statement $c < d$, $c \leq d$ is equivalent to ($c < d$ or $c = d$), and $p \neq q$ is a negation of $p = q$.

10. Statement A says that $x > 0$; statement B says that $x < -1$. What constraints on x establishes statement $\neg(A \wedge B)$? Illustrate this problem graphically. Do the same for statement $\neg(A \vee B)$.

11. For any number b, such that $b > 1$, it is the case that $b > 0$. Are statements $b < 1$ and $b > 0$ contradictory? Explain why by using the relationship between a statement and its inverse. Illustrate this problem graphically.
12. If x, a are numbers, then $x \geq a$ and $x \leq a$ if and only if $x = a$. What is then implied by $x \neq a$?
13. Inequality
$$(x - a) \cdot (x - b) < 0, \tag{1.12}$$
where a, b, and x are real numbers, implies
$$((x > a) \wedge (x < b)) \vee ((x < a) \wedge (x > b)). \tag{1.13}$$
Separately, inequality
$$(x - a) \cdot (x - b) > 0, \tag{1.14}$$
implies
$$((x > a) \wedge (x > b)) \vee ((x < a) \wedge (x < b)). \tag{1.15}$$
Since (1.14) is a negation of (1.12), we expect (1.15) to be a negation of (1.13) (see exercise 4). Show that this is indeed the case by negating (1.13).
14. Show that the statement $((x < a) \vee (x < b)) \wedge ((x > a) \vee (x > b))$ implies $(a < x < b) \vee (b < x < a)$. Illustrate this problem graphically.
15. Equation $x^2 + 2bx + c < 0$ is satisfied if and only if $b^2 > c$ and $x < -b + \sqrt{b^2 - c}$ and $x > -b - \sqrt{b^2 - c}$. Construct a negation of this statement. Make a plot of $x^2 + 2bx + c$ for $b = 1, c = 0$ as a function of x and illustrate both the original and the negated statement graphically.

For the following exercises select one or more correct answers. Explain why the other options are incorrect.

16. The entomologist Bénédict discovered a new species of spiders in Africa. He collected 131 individual spiders of this species, and all of them have 8 legs.

    (a) All spiders that belong to this species have 8 legs.
    (b) Some spiders that belong to this species have 8 legs.
    (c) Some spiders that belong to this species may have 6 legs.
    (d) All spiders have 8 legs, except the case when Bénédict's travel companion Hercules accidentally tore off 2 legs from a spider.

17. It is not true that Andy never rides a bike and never swims.
    (a) Andy sometimes rides a bike and swims.
    (b) Andy rode a bike, swam, or ran on at least one occasion.
    (c) Andy sometimes rides a bike, swims, or both.
18. It is not true that aliens ever landed in the Area 51 facility in Nevada.
    (a) Aliens do not exist or they do exist and landed elsewhere.
    (b) Aliens never landed in the Area 51 facility.
    (c) Aliens do not exist or they do exist but never landed in Area 51.
19. If entomologist Bénédict eats hot chicken wings, he immediately gets heartburn. For the last hour, he did not experience heartburn.
    (a) Bénédict did not have hot chicken wings recently.
    (b) The wings had a mild flavor.
    (c) Bénédict took a medicine against heartburn.
20. Astronauts arrive at a new planet and stay for 437 days there. They observe that it rains every day at sunrise.
    (a) It always rains at sunrise on this planet.
    (b) It rains at least on some days at sunrise.
    (c) The planet has an awful climate: it rains all the time.
21. We have learned that $x = 3$ and $x = 6$ are solutions of the equation $x^3 - 10x^2 + 27x - 18 = 0$.
    (a) $x = 6$ is a solution or $x = -1$ is a solution.
    (b) $x = 3$ and $x = 6$ are the only solutions of this equation.
    (c) $x = 1$ is not a solution.
22. Read about Pólya's conjecture on p. 22.
    (a) Pólya's conjecture is false.
    (b) Pólya's conjecture is almost always true.
    (c) Pólya's conjecture is rarely true.
23. We have learned that $x < 40$ implies that $x \leq 40$.
    (a) $0 < x < 40$.
    (b) $0 < x \leq 40$.
    (c) Statement $x < 40$ may or may not be true.
24. If Ada Lovelace gets a year-end bonus, then on December 31st, she will buy a brand new computer. Without this computer she cannot implement a new algorithm to compute the so-called Bernoulli numbers.

(a) Ada Lovelace bought a new computer, therefore the year-end bonus was not small.
(b) Ada Lovelace did not get a new computer on December 31st, therefore she did not get a year-end bonus.
(c) Ada Lovelace did not compute the Bernoulli numbers, therefore she did not receive her bonus.
(d) Being a daughter of a British peer, Ada Lovelace did not really need a year-end bonus.

25. If $x = b$, then $ax = ab$ for any a. (For selecting a correct statement or correct statements below, use your knowledge of algebra.)

(a) If $ax = ab$ for any a, then $x = b$.
(b) There exists such value of a that if $ax = ab$, then $x = b$.
(c) Statement $x = b$ does not follow from $ax = ab$.

26. For any b, the value $x = b$ is a solution of the following equation:

$$x - b = 0. \tag{1.16}$$

Which of the following statements is true:

(a) $x = 1$.
(b) The solution of this equation does not exist or is equal to b.
(c) There is such a value of b that the solution of equation (1.16) does not exist.

2

Numbers

2.1 Natural Numbers and Primes

Natural numbers are those that we learn first: $1, 2, 3, ...$ They are the starting point of our journey. In the following sections, we will build a better appreciation of how different types of numbers (natural, rational, real, and complex) form a harmonic, coherent system with unified and intelligible rules.

Here, we will rely on our intuitive understanding of what natural numbers are, just as mathematicians did for centuries before they came up with a rigorous definition. For natural numbers we define two operations: addition and multiplication. These operations have the following familiar properties:

1. Associativity: For any numbers m, n, k, it is the case that $(m+n)+k = m+(n+k)$ and $(m \cdot n) \cdot k = m \cdot (n \cdot k)$.
2. Commutativity: For any numbers m, n, it is the case that $m+n = n+m$ and $m \cdot n = n \cdot m$.
3. Distributivity: For any numbers m, n, k, it is the case that $m \cdot (n+k) = m \cdot n + m \cdot k$.
4. Unit element: For this property we have to introduce a new number – zero, which is denoted as 0. Then for any number n, it is the case that $n + 0 = n$ and $n \cdot 1 = n$.
5. Closure: A sum of two natural numbers is a natural number, and a product of two natural numbers is also a natural number. This property is sometimes missed because it appears obvious.

Can we flip back the closure property? Given a number, can we represent it as a sum or a product of two other numbers? For addition, the answer is yes: any number greater than 1 can be written as a sum of two other natural numbers. For multiplication, some natural numbers also can be represented as a product of two or more smaller factors. For example, $60 = 4 \cdot 15$. Such numbers are called composite. However, some other numbers (such as 13, 29, or 87,178,291,199) can be represented only as a single factor. Such numbers are called prime.

DOI: 10.1201/9781003456766-2

In 1742, the German mathematician Christian Goldbach suggested that every even natural number greater than 2 can be represented as a sum of two prime numbers. In many cases there are multiple ways to do that. For example,

$$
\begin{aligned}
&8 = 3 + 5,\\
&38 = 7 + 31 = 19 + 19,\\
&50 = 3 + 47 = 7 + 43 = 13 + 37 = 19 + 31,\\
&389,965,026,819,938 = 5,569 + 389,965,026,814,369.
\end{aligned}
\tag{2.1}
$$

This conjecture has been verified numerically for numbers up to 10^{17}, and there are strong arguments in favor of it being true for all even numbers, but it remains not proved.

Now that we have established the existence or prime and composite numbers, we can formulate the *fundamental theorem of arithmetic.* (It is also the very first theorem in his book):

Theorem 2.1 *Any integer greater than* 1 *can be represented uniquely as a product of prime numbers (ignoring the order of the factors).*

For example, $60 = 2 \cdot 2 \cdot 3 \cdot 5$. For the purposes of this theorem, such a product for a prime number is considered as one having a single factor: $29 = 29$.

If we parse the formulation of this theorem, we will see that that it contains two statements:

1. The desired product of primes exists, and
2. It is unique.

Here, we prove only the existence. (The proof of uniqueness is a bit more complicated.)

Proof

For this proof we use the method of strong mathematical induction. We start from the lowest number that this theorem applies to—that is, 2. *This is a prime number and therefore the theorem is true for it.*

Next, we assume that this theorem is true for all numbers from 2 *to* n *and prove it for number* $(n+1)$. *To do that we consider two cases:*

1. *Number* $(n+1)$ *is prime. Then the theorem is true for this number (in the sense that we described above).*
2. *Number* $(n+1)$ *is composite. By definition, we can represent it as a product of two smaller numbers:* $(n+1) = k \cdot m$. *However, due to the strong induction hypothesis, each of these factors can be represented*

as a product of some primes: $k = q_1 \cdot q_2 \cdot \ldots \cdot q_i$ *and* $m = p_1 \cdot p_2 \cdot \ldots \cdot p_j$. *Therefore, we get* $(n+1) = q_1 \cdot q_2 \cdot \ldots \cdot q_i \cdot p_1 \cdot p_2 \cdot \ldots \cdot p_j$.

Note that from the very definition of prime numbers it follows that every natural number can be represented as a product of *some* factors (as long as such a product for a prime number is considered as one having a single factor). The fundamental theorem of arithmetic says something different: any natural number is represented as a product of *primes*, and this representation is unique.

Take some natural number N and consider all numbers $2 \leq n \leq N$. According to the fundamental theorem of arithmetic, each number n can be represented as a product of prime factors. Let's now determine, how many of these representations have an even number of factors and how many have an odd number of factors. For example, for $N = 9$ we get

n	List of Factors	Even or odd number of factors
2	2	Odd
3	3	Odd
4	2,2	Even
5	5	Odd
6	2,3	Even
7	7	Odd
8	2,2,2	Odd
9	3,3	Even

Among the eight numbers n, five (2, 3, 5, and 7) produce an odd number of factors, and three (4, 6, and 9) produce an even number of factors. It looks like the cases when the number of factors is odd are helped by all the primes among values of n, when there is only one factor. Is this a general rule?

In 1919 the Hungarian mathematician George Pólya posed a conjecture that bears his name and that states that for any natural number N, at least 50% of numbers $2 \leq n \leq N$ produce an odd number of factors. This conjecture remained unresolved until 1958, when it was proved wrong. Today we have a counterexample for N= 906,150,257. For this number (and some numbers following it), more than 50% of numbers from 2 to N produce an even number of factors. This is not even helped by the fact that there are about 46 million primes from 2 to N. (We discuss counting primes on p. 148.)

Another great theorem deals with the number of primes. Is there a largest prime number? or is the number of primes infinite? Turns out, the former assumption is false, and the latter is true:

Theorem 2.2 *The number of primes is infinite.*

In this section we give two proofs of this theorem, and there will be a rather interesting third proof in section 9.15.

Proof

The classic proof by contradiction is a modification of a result by Euclid. Let's assume that there is only a finite number of primes. We denote these prime numbers as $p_1, p_2, ..., p_n$. *According to our assumption, this is the exhaustive list of primes; any other number is composite. Now, we construct a new number:*

$$N = p_1 \cdot p_2 \cdot ... \cdot p_n + 1. \tag{2.2}$$

Note that N *is not evenly divisible by any of the numbers* $p_1, p_2, ..., p_n$. *For example, if we try to divide* N *by* p_1, *we will get the quotient* $p_2 \cdot p_3 \cdot ... \cdot p_n$ *and the remainder*[1] *of 1.*

Since the number N *is not in our original exhaustive list of primes, it must be composite. By the fundamental theorem of arithmetic, it can be represented as a product of two or more primes:*

$$N = q_1 \cdot q_2 \cdot ... \cdot q_m. \tag{2.3}$$

We see that N *is divisible by a prime number* q_1. *Since our list* $p_1, p_2, ..., p_n$ *contains all prime numbers,* q_1 *must be one of them; for example* $q_1 = p_k$. *However, we have established earlier that* N *is not evenly divisible by* p_k. *Thus, we've reached a contradiction. This means that our initial premise was false and that the number of primes must be infinite.*

> Given prime factors, it is easy to compute their product, but given a composite number, it is can be difficult to find its prime factors. Often, we have to try different prime numbers and see if each is a factor. This fact, and the relatively high frequency of occurrence of prime numbers form the basis for modern data encryption algorithms.

For the second proof we need the following

Lemma 2.1 *if both* P *and* R *are multiples of* Q *(that is,* $P = kQ$ *and* $R = lQ$*, where* k, l *are integers), then* $P - R$ *is also a multiple of* Q.

Proof

If $P = kQ$ *and* $R = lQ$*, then* $P - R = (k - l)Q$. *Therefore,* $P - R$ *is a multiple of* Q.

With this lemma, the second proof of theorem 2.2 is as follows.

Proof

This proof is also by contradiction. Similarly to Euclid's proof, we assume that there is a finite number of primes, $p_1, p_2, \ldots, p_n$. *Next, we construct numbers*

[1] There is another theorem that states that for any pair of integers, the quotient and the remainder exist and are unique.

$N = p_1 \cdot p_2 \cdot \ldots \cdot p_n$ and $M = N - 1 = p_1 \cdot p_2 \cdot \ldots \cdot p_n - 1$. Since the original list of primes is complete, M must be composite. Therefore, it is a product of some primes from the original list. We select one of them and denote it p', so that M is a multiple of p'. We also know that N is a multiple of p', since N is the product of the exhaustive list of primes. From the lemma it follows that $N - M$ is a multiple of p'. However, $N - M = 1$, which cannot be a multiple of p', producing a contradiction.

There is no known formula that would predict all the prime numbers, in spite of a lot of effort to find one. Still, there are various patterns that can be used to "predict" primes with various degrees of robustness and utility. Three of them are listed below:

1. *Mersenne numbers* are defined as $2^p - 1$, where p is a prime. There is a surprisingly high number of primes among the Mersenne numbers; they are called Mersenne primes. The first Mersenne primes are given by exponents $p = 2, 3, 5, 7, 13, 17, 19, 31, 61$, and 89. There are very large Mersenne primes that are produced by $p = 43{,}112{,}609$, with $2^p - 1$ having 12,978,188 digits, or $p = 82{,}589{,}933$, with $2^p - 1$ having 24,862,048 digits.
2. *Leonhard Euler's numbers* are defined as

$$N = k^2 - k + n. \tag{2.4}$$

   There are such values of n, that equation (2.4) produces $n - 1$ primes for all $1 \leq k < n$. The values of n that produce primes this way are 2, 3, 5, 11, 17 and 41. (Note that all of them are also primes.) For example, for $n = 3$ we get two primes: 3 (for $k = 1$) and 5 (for $k = 2$). For $n = 41$ we get 40 primes: 41, 43, 47, 53, 61, 71, 83, 97, 113, 131, 151, 173, 197, 223, 251, 281, 313, 347, 383, 421, 461, 503, 547, 593, 641, 691, 743, 797, 853, 911, 971, 1033, 1097, 1163, 1231, 1301, 1373, 1447, 1523, 1601, 1847, 1933, 2111, 2203, 2297, 2393, 2591, 2693, and 2797.
3. There is a theorem proving the existence of a real number A, such that expression $\lfloor A^{3^n} \rfloor$ is prime for every natural n. (Here $\lfloor x \rfloor$ means "the integer part of x." For example $\lfloor 1.1 \rfloor = 1$, and $\lfloor 7.9 \rfloor = 7$.) At the first glance this looks like a great way to produce primes, but a closer look into this theorem shows that the only known way to determine number A requires us to already know the values of the primes that would then be predicted by this formula, so this strange result remains just a curiosity.

2.2 Integers

Operations of addition and multiplication are often used to solve various practical problems. For example, in some application we may need to determine a natural number N that satisfies the following equation:

$$N + 5 = 7. \tag{2.5}$$

For equations of this type we define the *inverse operation*, called subtraction:

$$N = 7 - 5. \tag{2.6}$$

Once we have learned how to do subtraction, we conclude that $N = 2$.

Unfortunately, the subtraction operation does not have the closure property in the realm of natural numbers: sometimes it works, and sometimes it does not. For example, there is no natural number that corresponds to the expression $5 - 7$.

To fix that problem, we introduce zero and negative numbers. Together with natural numbers, they form what are called the integers. For example, $5, 0, -11780$, and $-8764356 \cdot 10^{765765}$ are integers. Then any equation like $x + n = m$ will have a solution $x = m - n$. For example, if $x + 1 = 0$, then $x = -1$.

We extend all the properties of natural numbers to integers. It is best to do so in a way that preserves all the properties of multiplication and addition. One consequence of this is that

$$n \cdot 0 = 0, \tag{2.7}$$

for any n. Indeed, let's distribute the following product:

$$(m + 0) \cdot n = m \cdot n + n \cdot 0. \tag{2.8}$$

From property 4 on p. 20, we get $m \cdot n$ in the left-hand side, then

$$m \cdot n = m \cdot n + n \cdot 0, \tag{2.9}$$

which implies equation (2.7).

Another consequence of extending all the properties of natural numbers to integers is the law of signs for multiplying negative numbers. Consider the distributive property:

$$m \cdot (n + k) = m \cdot n + m \cdot k. \tag{2.10}$$

What happens if we use $m = -1, n = 1, k = -1$? The last equation would take the form

$$-1 \cdot (1 + (-1)) = -1 \cdot 1 + (-1) \cdot (-1). \tag{2.11}$$

From the way we introduced negative numbers we know that the equation

$$x + 1 = 0, \tag{2.12}$$

implies that $x = -1$. Therefore, we must have $1 + (-1) = 0$. Then the left-hand side of equation (2.11) produces $-1 \cdot 0 = 0$ because of equation (2.7). We get

$$0 = -1 \cdot 1 + (-1) \cdot (-1). \tag{2.13}$$

The term $-1 \cdot 1$ in the right-hand side of this equation must be equal to -1 because of property 4 on p. 20. Then

$$0 = -1 + (-1) \cdot (-1). \tag{2.14}$$

We see that the right-hand side must be equal to zero, but the only way to ensure that is to have $(-1)\cdot(-1) = 1$. We can generalize this result: a product of two negative numbers is positive. This rule of signs ensures consistency for the rules of addition and multiplication for integer numbers.

The next useful concept deals with even and odd numbers. We define even integers as those that are divisible by 2; this means that they can be presented as $n = 2k$. Odd ones are those that are not divisible by 2; they can be presented as $m = 2l + 1$. Even and odd numbers have the following properties:

Theorem 2.3 *Operations with odd and even numbers have the following properties:*

1. *A sum of two even numbers is even.*
2. *A sum of two odd numbers is even.*
3. *A sum of an even and an odd number is odd.*
4. *A product of two even numbers is even.*
5. *A product of two odd numbers is odd.*
6. *A product of an even and an odd number is even.*

Proof

All the statements of this theorem are proved by using $N_e = 2k$ for the even number, $N_o = 2l + 1$ for the odd number, and constructing the corresponding sums and products. For example, consider the product of an even and an odd number:

$$\begin{aligned} N_e N_o &= 2k(2l+1), \\ &= 2(k(2l+1)). \end{aligned} \tag{2.15}$$

Since $k(2l + 1)$ is a natural number due to the closure property, $2(k(2l + 1))$ is even.

Negative numbers first appeared in the Chinese treatise *Nine Chapters on the Mathematical Art.* An extensive commentary to this treatise, with rules for the addition and subtraction of negative numbers, was written by Liu Hui in 263 CE. In later centuries negative numbers were considered in Indian and Islamic manuscripts.

However, the concept of negative numbers remained unknown in Europe for a long time. In the 16th century, Gerolamo Cardano went to great lengths trying to avoid using negative numbers when solving cubic equations.

2.3 Rational Numbers

Sometimes an application requires us to solve an equation, where the unknown is multiplied by a number. An example of such an equation is as follows:

$$M \cdot 5 = 10. \tag{2.16}$$

After thinking about it, we determine that $M = 2$. Since we are no longer limited to natural numbers, we can even solve equations like $M \cdot 5 = -10$. However, very soon we arrive at a similar problem: not all equations of this kind can be solved within the realm of integers. For example, there is no integer that satisfies equation

$$K \cdot 2 = 3. \tag{2.17}$$

This prompts a new extension: we define new objects, called rational numbers. The notation for these new numbers is fractions. In this notation, the solution of equation (2.17) is defined as

$$K = \frac{3}{2}. \tag{2.18}$$

For these rational numbers we formulate the following properties:

$$\begin{aligned} &\frac{a}{b} + \frac{c}{d} = \frac{ad + bc}{bd}, \\ &\frac{a}{b} \cdot \frac{c}{d} = \frac{ac}{bd}, \\ &\frac{a}{1} = a. \end{aligned} \tag{2.19}$$

We also state that two rational numbers $\frac{a}{b}$ and $\frac{c}{d}$ are equal if and only if $ad = cb$. From this we get

$$\frac{ac}{bc} = \frac{a}{b}. \tag{2.20}$$

Equation (2.20) allows us to define a rational number in the *canonical form*. If the numerator and denominator have common factors, we can "cancel" them to arrive at lowest possible values for both. For example, $12/15 = (3 \cdot 4)/(3 \cdot 5) = 4/5$.
Equations (2.19) are very important: they are formulated to guarantee preservation of all the properties for addition and multiplication. Indeed, the closure for these operations is directly evident from equations (2.19). For other properties we prove the following theorems:

Theorem 2.4 *Addition and multiplication of rational numbers is commutative:*

$$\begin{aligned} \frac{m}{n} + \frac{k}{l} &= \frac{k}{l} + \frac{m}{n}, \\ \frac{m}{n} \cdot \frac{k}{l} &= \frac{k}{l} \cdot \frac{m}{n}. \end{aligned} \tag{2.21}$$

Proof

1. *Consider both sides of the first equation in (2.21). By definition, we have*
$$\frac{m}{n} + \frac{k}{l} = \frac{ml + kn}{nl}, \tag{2.22}$$
and
$$\frac{k}{l} + \frac{m}{n} = \frac{kn + ml}{ln}. \tag{2.23}$$
A comparison of the right-hand sides of the last two equations shows that commutativity for addition holds.
2. *Similarly, we compare*
$$\frac{m}{n} \cdot \frac{k}{l} = \frac{mk}{nl}, \tag{2.24}$$
and
$$\frac{k}{l} \cdot \frac{m}{n} = \frac{km}{ln}. \tag{2.25}$$
to prove that commutativity for multiplication holds as well.

Theorem 2.5 *Addition and multiplication of rational numbers is associative:*

$$\begin{aligned} \left(\frac{m}{n} + \frac{k}{l}\right) + \frac{p}{q} &= \frac{m}{n} + \left(\frac{k}{l} + \frac{p}{q}\right), \\ \left(\frac{m}{n} \cdot \frac{k}{l}\right) \cdot \frac{p}{q} &= \frac{m}{n} \cdot \left(\frac{k}{l} \cdot \frac{p}{q}\right). \end{aligned} \tag{2.26}$$

Proof

1. *We prove associativity for addition first. Consider both sides of the first equation in (2.26). By definition, we have*

$$\begin{aligned} \left(\frac{m}{n}+\frac{k}{l}\right)+\frac{p}{q} &= \frac{ml+kn}{nl}+\frac{p}{q} \\ &= \frac{(ml+kn)q+pnl}{nlq} \\ &= \frac{mlq+knq+pnl}{nlq} \end{aligned} \tag{2.27}$$

and

$$\begin{aligned} \frac{m}{n}+\left(\frac{k}{l}+\frac{p}{q}\right) &= \frac{m}{n}+\frac{kq+pl}{ql} \\ &= \frac{mql+(kq+pl)n}{nlq} \\ &= \frac{mlq+knq+pnl}{nlq}. \end{aligned} \tag{2.28}$$

A comparison of the final expressions in equations (2.27) and (2.28) shows that associativity holds.

2. *The proof of associativity for multiplication is simpler:*

$$\left(\frac{m}{n}\cdot\frac{k}{l}\right)\cdot\frac{p}{q}=\frac{mkp}{nlq}=\frac{m}{n}\cdot\left(\frac{k}{l}\cdot\frac{p}{q}\right). \tag{2.29}$$

Theorem 2.6 *Rational numbers have the distributive property:*

$$\frac{p}{q}\cdot\left(\frac{m}{n}+\frac{k}{l}\right)=\frac{p}{q}\cdot\frac{m}{n}+\frac{p}{q}\cdot\frac{k}{l}. \tag{2.30}$$

Proof

We consider the left and the right-hand sides separately:

$$\begin{aligned} \frac{p}{q}\cdot\left(\frac{m}{n}+\frac{k}{l}\right) &= \frac{p}{q}\cdot\frac{ml+kn}{nl} \\ &= \frac{p(ml+kn)}{qnl} \\ &= \frac{pml+pkn}{qnl} \end{aligned} \tag{2.31}$$

and

$$\frac{p}{q}\cdot\frac{m}{n}+\frac{p}{q}\cdot\frac{k}{l}=\frac{pm}{qn}+\frac{pk}{ql}$$
$$=\frac{pmql+pkqn}{qnql}$$
$$=\frac{q(pml+pkn)}{qnql} \tag{2.32}$$
$$=\frac{pml+pkn}{qnl}.$$

A comparison of the final expressions in equations (2.31) and (2.32) proves the theorem.

The last equation in (2.19) shows that any integer can be viewed as a rational number. The result of arithmetic operations for adding or multiplying integers is the same, whether we perform them in a straightforward way, as we usually manipulate integers, or by using the rules for rational numbers:

$$a+b=\frac{a}{1}+\frac{b}{1},$$
$$=\frac{a\cdot 1+b\cdot 1}{1\cdot 1} \tag{2.33}$$
$$=\frac{a+b}{1}$$
$$=a+b.$$

and

$$a\cdot b=\frac{a}{1}\cdot\frac{b}{1},$$
$$=\frac{a\cdot b}{1\cdot 1} \tag{2.34}$$
$$=\frac{a\cdot b}{1}$$
$$=a\cdot b.$$

This consistency is important: we can use different types of numbers in a unified way and we do not have to make separate rules that apply to various special cases.

The definition of rational numbers has one limitation: there are no rational numbers that are given by $\frac{n}{0}$ for any n. Can we use the same trick as before and define a number that is equal to $\frac{n}{0}$? Unfortunately, the answer is "no." Indeed, suppose that we have defined some special number ρ that is equal to $\frac{1}{0}$. By definition, that would mean that

$$\rho\cdot 0=1. \tag{2.35}$$

Let's multiply this equation by some other number, for example, by 2. Because of the associativity of multiplication, we get:

$$\rho \cdot (0 \cdot 2) = \rho \cdot 0 = 2. \tag{2.36}$$

We see that multiplication by ρ does not produce a unique result. For this reason, we do not introduce any number like ρ but simply forbid all divisions by zero.

When doing computations with rational numbers we often use helpful shortcuts, which are ultimately derived from the basics. The following theorem shows how to compute a ratio of two numbers.

Theorem 2.7 *For any two rational numbers m/n and k/l, such that $k \neq 0$, it is the case that*

$$\frac{\left(\frac{m}{n}\right)}{\left(\frac{k}{l}\right)} = \frac{ml}{kn}. \tag{2.37}$$

Proof

We multiply equation (2.37) by the denominator in the left-hand side to get

$$\frac{m}{n} = \frac{k}{l} \cdot \frac{ml}{kn}. \tag{2.38}$$

The right-hand side simplifies to m/n, which proves the theorem.

From the above, it looks like we can operate completely within the domain of rational numbers, with no surprises. Make no mistake, this is important. It would be a bad thing if we added two numbers and got something that is not a number as a result. Luckily, for the four arithmetic operations, there is no need to go beyond rational numbers. That was what ancient Greeks thought before they asked a question: what is the length of the diagonal of a square? We will answer this question in section 2.5.

Rational numbers do not have finite-size gaps. This is formalized using the following theorem.

Theorem 2.8 *There is a rational number between any two different rational numbers.*

Proof

This is a constructive proof. For any two rational numbers r_1 and r_2, we construct $r = (r_1 + r_2)/2$. (A proof that $r = (r_1 + r_2)/2$ is in between r_1 and r_2 is given in section 3.3.2.)

Therefore, any gap between two rational numbers always has a rational number in it, which creates two smaller gaps. We can continue this process indefinitely, which means that gaps between rational numbers are infinitely small.

2.4 The Decimal Representation

2.4.1 Decimal Representation of Integers

Today we represent integers by using the place-value system: the contribution of a digit to the value of a number is the value of the digit multiplied by a factor that is determined by the position of that digit. For example, the value of number the 729 comprises the following: the last digit contributes 9, the middle digit contributes $2 \cdot 10 = 20$, and the first digit contributes $7 \cdot 100 = 700$. The sum of these numbers produces a desired value that is given by the sequence of digits 729. Can we be sure that any natural number can be represented using this system? The answer to this question is given by the following theorem.

Fibonacci, also known as Leonardo of Pisa, was one of the greatest mathematicians of the Middle Ages in Europe. In chapter 10 we will explore the eponymous Fibonacci numbers that he introduced as an example in his major mathematical treatise. That book popularized the place-value system that we continue using today. This system originated in India and then spread to the Middle East and parts of the Islamic World. Today it is often called Arabic or Hindu-Arabic. It is superior to the Roman numerals that were in use in Europe before that.

It is no accident that Fibonacci recognized the convenience of Hindu-Arabic numerals: as a boy, he accompanied travels of his merchant father all over the Mediterranean, including North Africa, where the place-value system has already been in use. In his book, Fibonacci gives credit to the culture where that system had originated: he calls it "the method of the Indians."

Theorem 2.9 *Any natural number N can be represented using the place-value system.*

Proof

This proof is by strong induction.

1. *For $N = 1$ we get $N = 1$.*
2. *We assume that the theorem is true for N. For $N + 1$ we select the largest element of sequence $1, 10, 100, \ldots$ that does not exceed $N + 1$. Let's denote that element as K. Then there are two possibilities:*
   - *(a) $K = N+1$. Then the statement of the theorem is true. Note that this corresponds to the case when the first digit of the number is 1 and other digits are zeros (such as 1000).*

(b) $K < N+1$. *Number K will still have the first digit equal to* 1 *and the remaining digits equal to* 0. *We compute a positive integer $L = (N+1) - K$. Number L will differ from number $N+1$ only by the first digit, which will be reduced by* 1 *(as in* $729 - 100 = 629$ *or* $129 - 100 = 029 = 29$*). According to our assumption, L can be represented by the place-value system. Then number $N+1$ will have a similar representation, except the first digit must be incremented by* 1.

Moreover, such a representation is unique, except for leading zeros, which are always omitted in practice. A proof of uniqueness is the subject of exercise 29 for this chapter.

This seemingly trivial theorem can be generalized to other ways to represent integers. In section 10.7 we will show that representing integers as a sum of Fibonacci numbers has curious properties.

2.4.2 Decimal Representation of Rational Numbers

We commonly represent rational numbers as decimals. For example, $\frac{3}{2} = 1.5$, which is just a notation for $1 + \frac{5}{10}$. Some rational numbers require infinitely long decimal representation. For example, $\frac{1}{3} = 0.3333...$

A decimal representation is not always unique. Consider $0.9999\ldots = 0.\overline{9}$. Let's denote this number as a:

$$a = 0.9999\ldots \tag{2.39}$$

Then

$$10a = 9.999\ldots \tag{2.40}$$

Let's subtract equation (2.39) from equation (2.40):

$$10a - a = 9.999\ldots - 0.9999\ldots \tag{2.41}$$

We get:

$$9a = 9, \tag{2.42}$$

which means that $a = 1$, or:

$$0.9999\ldots = 1. \tag{2.43}$$

It can't be any other way!

The decimal representations of some rational numbers contain infinitely many digits. How does that work? Consider a decimal representation of 10/7:

$$\frac{10}{7} = 1.428571428571428571428571... \tag{2.44}$$

We notice that this decimal sequence seems to be *periodic*—that is, the first 24 digits after the decimal point show that the numbers 428571 repeat four times.

Let us assume that this pattern of repeating numbers continues indefinitely. Then the right-hand side can be written as

$$\frac{10}{7} = 1 + 428571 \cdot (10^{-6} + 10^{-12} + 10^{-18} + ...). \tag{2.45}$$

Below (see theorem 9.18) we prove that this infinite sum is given by

$$\frac{10}{7} = 1 + 428571 \frac{10^{-6}}{1 - 10^{-6}}. \tag{2.46}$$

This is simplified to

$$\frac{10}{7} = 1 + \frac{428571}{999999}. \tag{2.47}$$

This is a rational number. Indeed, if we compute the common denominator, we get:

$$1 + \frac{428571}{999999} = \frac{1428570}{999999}. \tag{2.48}$$

This is a ratio of two integers, and therefore is a rational number by definition.

Why does it not look like the original 10/7? Can we find common divisors for the numerator and the denominator in the above fraction? Yes, we can:

$$\begin{aligned} 1428571 &= 2 \cdot 3^3 \cdot 5 \cdot 11 \cdot 13 \cdot 37, \\ 999999 &= 3^3 \cdot 7 \cdot 11 \cdot 13 \cdot 37. \end{aligned} \tag{2.49}$$

After canceling the common divisors we do get

$$\frac{1428571}{999999} = \frac{2 \cdot 3^3 \cdot 5 \cdot 11 \cdot 13 \cdot 37}{3^3 \cdot 7 \cdot 11 \cdot 13 \cdot 37} = \frac{10}{7}. \tag{2.50}$$

This example shows why any periodic infinite decimal representation produces a rational number. We represent it as an infinite sum, which can be transformed into a ratio of integers. This can be turned into a formal proof, of course. Canceling all common divisors produces the canonical form of a rational number. In the above example, 10/7 is a canonical form, and 1428571/999999 is not.

> Amazingly, if we take a random rational number in the canonical form, the probability that its denominator is even is equal to 1/3.

It is a bit more difficult to prove the converse statement: that any rational number produces a finite or periodic decimal representation. To outline the idea behind this proof, let's consider the mechanics of computing the decimal representation of 10/7. First, we compute the quotient—that is, the integer part:

$$\lfloor \frac{10}{7} \rfloor = 1, \tag{2.51}$$

where $\lfloor . \rfloor$ means "the integer part." Then

$$\frac{10}{7} = 1 + \frac{3}{7}. \tag{2.52}$$

This gives us the first digit of the decimal representation—that is, 1. The fraction in the right-hand side is less than 1, but we can factor out a 0.1:

$$\begin{aligned} \frac{10}{7} &= 1 + \frac{3}{7} \\ &= 1 + 0.1 \cdot \frac{30}{7}. \end{aligned} \tag{2.53}$$

Now fraction 30/7 is greater than 1 and we repeat the quotient computation to produce the second digit of the decimal representation:

$$\lfloor \frac{30}{7} \rfloor = 4. \tag{2.54}$$

The remainder of this division is 2. We obtain two digits (note the 0.1 that was factored out before):

$$\frac{10}{7} = 1 + 0.4 + 0.1 \cdot \frac{2}{7}. \tag{2.55}$$

We repeat the same trick to get the third digit:

$$\frac{10}{7} = 1 + 0.4 + 0.01 \cdot \frac{20}{7}, \tag{2.56}$$

where we observe that

$$\lfloor \frac{20}{7} \rfloor = 2, \tag{2.57}$$

and so on. The important part is that at each step the numerator originally is a positive number that is less than 7, and, after multiplying it by 10, less than 70. This means that there is a finite number of options for the numerator. Therefore, if we perform these steps a large enough number of times, at some point we are bound to get a number in the numerator that we have encountered before. Starting from that, the whole sequence will repeat as well. We will get a *repeating pattern* of digits.

Sometimes the process of computing digits in the decimal representation of a noninteger rational number does not produce a remainder at one of the steps. Then the next digit and all other digits after that are zeroes. Such rational numbers, for example 1.5 or −3.00097, have a finite number of digits in their decimal representation (if we do not show trailing zeroes). Let's see when this is the case. As we compute successive digits, we continue multiplying the numerator by 10, which effectively continues contributing factors 2 and 5 to it. For the numerator to become evenly divisible by the denominator, the latter should be a product of a some combination of 2 and 5. Then at some step the numerator will accumulate enough twos and fives to become evenly divisible by the denominator. This gives us the criterion for a finite decimal representation of a rational number: its denominator must be a product of twos and fives. For example, $37/125 = 0.296$ and $37/128 = 0.2890625$. The finite number of digits in these examples is made possible by the fact that $125 = 5^3$ and $128 = 2^7$.

2.5 Irrational Numbers

We have already seen how a desire to have fully and consistently defined inverse operations prompts us to expand the set of numbers to include first zero and negative integers, and then rational numbers.

Today, we often use more operations than the basic four (addition, subtraction, multiplication, and division), such as exponentiation. For these operations we also want to define inverses. For example, it follows from the Pythagorean theorem that the square of the length of a diagonal of a unit square is equal to 2. Let's try to find a number whose square is equal to 2. (We will denote such a number as $\sqrt{2}$.) So far, we have been using rational numbers, so our initial search for $\sqrt{2}$ would be in that realm. Turns out, such a rational number does not exist.

For a long time ancient Greeks believed that the length of all line segments in any geometrical figure can be expressed as an integer multiple of some small unit of measure. If that were the case, the ratio of lengths of any two line segments would be a rational number. Such segments are called *commensurable.*

That the diagonal of a square is *incommensurable* with the side is believed to be discovered by Hippasus, a Pythagorean philosopher. Today we would say that the ratio of the length of the diagonal to the length of a side is an irrational number.

However, written sources about the role of Hippasus in discovering incommensurability are scant. The first explicit reference to incommensurability we find in Plato's dialogue *Theaetetus*, which in turn refers to Theodorus, a mathematician in the 5th century BCE.

Theorem 2.10 *There is no rational number whose square is equal to* 2.

Here are two proofs of this theorem. Both are by contradiction and both are quite elegant.

Proof

We assume that $\sqrt{2}$ is a rational number, which means that it can be represented as

$$\sqrt{2} = \frac{l}{k}. \tag{2.58}$$

We multiply that equation by k and square the result:

$$2k^2 = l^2. \tag{2.59}$$

According to the fundamental theorem of arithmetic, both k and l can be represented as products of prime factors:

$$\begin{aligned} k &= p_1 \cdot p_2 \cdot \ldots \cdot p_m, \\ l &= q_1 \cdot q_2 \cdot \ldots \cdot q_n. \end{aligned} \tag{2.60}$$

A factor of 2 can be a part of either representation. Let's suppose that a factor of 2 appears i times in $p_1 \cdot p_2 \cdot \ldots \cdot p_m$, and j times in $q_1 \cdot q_2 \cdot \ldots \cdot q_n$. Then a factor of 2 will appear $2i$ times in the representation of k^2 and $2j$ times in the representation of l^2.

The left-hand side of equation (2.59) is itself an integer. We denote this integer as M. It can be represented by its own product of prime factors. The factor 2 will appear there $2i+1$ times. The right-hand side of equation (2.59) is the same integer M. In its prime factor representation, factor 2 appears $2j$ times. But then factor 2 appears $2i+1$ times in the left-hand side of equation (2.59) and $2j$ times in the right-hand side, leading to

$$2i + 1 = 2j. \tag{2.61}$$

This cannot be true because we now have an odd number equal to an even number. We have arrived at a contradiction. This means that our initial assumption about the rationality of $\sqrt{2}$ was false.

YBC 7289 is a Babylonian clay tablet (figure 2.1) that contains an excellent approximation to the square root of 2. The diagonal of the square is labeled with two sexagesimal (that is, base 60) numbers. The first of these two numbers is 1; 24, 51, 10. It represents the number 305470/216000 ≈ 1.414212963. Compare this to $\sqrt{2} \approx 1.41421356237$.

The second proof that $\sqrt{2}$ is not a rational number is as follows.

Proof

This proof is also by contradiction, but does not use the fundamental theorem of arithmetic.

We assume that there exists a rational number r, such that $r^2 = 2$. It is easy to see that this number must be greater than 1 and less than 2. If r is rational, there exist positive integer numbers N_j, such that $N_j r$ are integers.

FIGURE 2.1
Babylonian clay tablet YBC-7289. (Yale Babilonian Collection. A modified version of a photo by Bill Casselman, with cuneiform markings emphasized.)

(Indeed, if $r = l/k$, where l, k are integers, numbers N_j are multiples of k.) Among these numbers N_j, there is the smallest one, which we denote as n. Next, we construct a new positive integer number m:

$$m = n\sqrt{2} - n. \tag{2.62}$$

If we compute the value of $m\sqrt{2}$, we shall see that it is also a positive integer. Indeed,

$$m\sqrt{2} = 2n - n\sqrt{2}. \tag{2.63}$$

The right-hand side is an integer because it is the difference of two integers, $2n$ and $n\sqrt{2}$. Since $1 < \sqrt{2} < 2$, we can see from equation (2.62) that m is positive and smaller than n. Then n is not the smallest of the numbers N_j, which is a contradiction.

This proof has some similarities with a proof of theorem 10.3 on p. 250.

To enable additional computations, such as the computation of radicals or the computation of the circumference of a circle given its radius, we introduce a new type of numbers called *irrational.* We can generalize theorem 2.10: take an integer n and seek $r = \sqrt{n}$. Then r is either an integer or an irrational number.

Irrational numbers have the following properties:

Theorem 2.11 *An irrational number has aperiodic decimal representation.*

Proof
The proof is by contrapositive. Indeed, a finite or periodic decimal representation corresponds to a rational number (section 2.4.2).

Theorem 2.12 *An infinite aperiodic decimal corresponds to an irrational number.*

Proof
Again, the proof is by contrapositive. We use section 2.4.2 where we saw that a rational number has a finite or periodic decimal representation.

2.5.1 Mixing Rational and Irrational Numbers

Rational and irrational numbers comprise so-called *real numbers.* For real numbers we can define all the familiar operations that possess the usual properties: commutativity, associativity, distributivity, and closure. Depending on the types of numbers we use, the result of arithmetic operations may be a rational or an irrational number. This is the subject of the following theorems.

> The German mathematician Richard Dedekind provided an ingenious rigorous construction of real numbers, which enables the extension of all arithmetic operations (addition, subtraction, multiplication, division, and exponentiation) from rationals to reals. This contribution is known as the *Dedekind cut.*
>
> In this book, we do not define operations with irrational numbers but simply assume that they work similarly to those with rational numbers.

Theorem 2.13 *The sum of a rational and an irrational number is an irrational number.*

Proof
The proof is by contradiction. We assume that

$$r_1 + w = r_2, \tag{2.64}$$

where r_1, r_2 are rational numbers, and w is irrational. Then

$$w = r_2 - r_1. \tag{2.65}$$

The right-hand side of this equation is a rational number, and the left-hand side is an irrational number, which is a contradiction that proves the theorem.

Theorem 2.14 *The product of a nonzero rational number and an irrational number is an irrational number.*

Proof
The proof is analogous to that of the previous theorem.

However, the sum of two irrational numbers can be either a rational or an irrational number. Indeed, from theorem 2.13 we know that a sum of a rational and irrational numbers is irrational. Algebraically this is written as

$$r + w = v, \tag{2.66}$$

where r is a rational number and w, v are irrational numbers. Then

$$v - w = r, \tag{2.67}$$

which proves that a sum of irrational numbers v and $-w$ can be a rational number. To prove that a sum of two irrational numbers can be an irrational number, we consider

$$s = \sqrt{2} + \sqrt{2} = 2\sqrt{2}. \tag{2.68}$$

Here the right-hand side is irrational per theorem 2.14. Similar arguments work for the product of two irrational numbers.

A very interesting proof deals with raising an irrational number to an irrational power.

Theorem 2.15 *An irrational number raised to an irrational power*[2] *can be a rational number.*

Proof
To prove an existence statement we need to find just one example. Let's consider the number $a = \left(\sqrt{2}\right)^{\sqrt{2}}$. There are only two possibilities: this number is either rational or irrational. We consider both of them in turn:

1. *Number a is rational. Then, we proved the theorem: an irrational number $(\sqrt{2})$, raised to an irrational power $(\sqrt{2})$ is rational.*
2. *Number a is irrational. Now let's compute $a^{\sqrt{2}}$:*

$$\begin{aligned} a^{\sqrt{2}} &= \left(\left(\sqrt{2}\right)^{\sqrt{2}}\right)^{\sqrt{2}} \\ &= \left(\sqrt{2}\right)^{\sqrt{2}\cdot\sqrt{2}} \\ &= \left(\sqrt{2}\right)^{2} \\ &= 2. \end{aligned} \tag{2.69}$$

According to our assumption, a is irrational. Raised to the irrational power of $\sqrt{2}$, it produces a 2, which is a rational number.

[2] We will talk about how to raise a number to an irrational power in sections 6.4 and 9.7.

Here is the best part: we still do not know if a is irrational or not, but either way, we know that it is possible to raise an irrational number to an irrational power to produce a rational number! This is a proof of existence in its purest form: we proved the existence without pointing to the object in question.

Theorem 2.16 *Between any two different rational numbers there is an irrational number.*

Proof

The proof is by construction. We start from two rational numbers, r_1 and r_2. Without loss of generality we assume that $r_2 > r_1$. Next, we construct number $s = r_1 + (r_2 - r_1)/\sqrt{2}$. From theorems 2.13 and 2.14 we conclude that s is irrational. We also see that $r_1 < s < r_2$. Indeed,

$$\begin{aligned} s - r_1 &= \frac{r_2 - r_1}{\sqrt{2}} > 0, \\ s - r_2 &= r_1 + \frac{r_2 - r_1}{\sqrt{2}} - r_2 \\ &= (r_2 - r_1)\left(\frac{1}{\sqrt{2}} - 1\right) < 0. \end{aligned} \tag{2.70}$$

Theorem 2.17 *Between any two different irrational numbers there is a rational number.*

Proof

A proof is by construction. We consider two irrational numbers w, v and compute $u = (w + v)/2$. If u is a rational number, we are done. If u is irrational, we consider its decimal representation and truncate it after n digits to obtain a rational number u_n. We select n large enough for $u - u_n < |w + v|/2$. Then u_n is guaranteed to be in the interval (w, v).

The next theorem shows what happens when a mix of integers and an irrational number is raised to an integer power.

Theorem 2.18 *Any integer power of $m + n\sqrt{k}$, where m, n, k are integers, can be represented in the form*

$$(m + n\sqrt{k})^N = l + j\sqrt{k}, \tag{2.71}$$

where l, j are integers.

Proof

The proof is by induction over N.

1. *For $N = 1$ the statement of the theorem turns to an identity if we set $l = m, j = n$.*

2. *Suppose that the theorem is true for N. Then, we get for $N+1$:*

$$(m+n\sqrt{k})^{N+1} = (l+j\sqrt{k})(m+n\sqrt{k}) \\ = (ml+jnk)+(jm+ln)\sqrt{k}. \tag{2.72}$$

Since $ml+jnk$ and $jm+ln$ are integers, equation (2.72) proves the theorem.

We will use this theorem to prove some amazing properties of the so-called Pell numbers in section 10.11.2. Theorem 2.18 also can be used for a proof of the statement that we made on p. 38: if an integer k is not a square of some other integer, then $\sqrt{k}$ is an irrational number. First, we prove the following

Lemma 2.2 *Consider numbers $K_{j,l}(p/q) = jp/q+l$ that are formed by multiplying a noninteger rational number p/q by some integer j and adding integer l for different possible values of j, l. Then each of the numbers $K_{j,l}$ is either an integer or differs from the nearest integer by at least $1/|q|$.*

Proof

Consider the absolute value of the difference between $jp/q+l$ and the nearest integer M:

$$\left| j\frac{p}{q}+l-M \right| = \frac{|jp+lq-Mq|}{|q|}. \tag{2.73}$$

If $jp/q+l$ is a noninteger, then $jp+lq \neq Mq$, and the value in equation (2.73) is nonzero. But then the numerator has the minimum value of 1, which proves the lemma.

Now, we can proceed to proving that $\sqrt{k}$ is either an integer or an irrational number.

Proof

If $\sqrt{k}$ is an integer, we are done. If it is not an integer, we have to prove that it is irrational. The proof is by contradiction. We compute the difference between $\sqrt{k}$ and the nearest integer M and raise it to power N. That is, we define D_N as

$$D_N = \left| \sqrt{k} - M \right|^N. \tag{2.74}$$

Since $D_1 < 1$, the numbers D_N decrease with N and become arbitrarily small[3] *for large values of N. According to theorem 2.18, we can represent D_N as*

$$D_N = \left| \left(\sqrt{k}-M\right)^N \right| \\ = \left| l + j\sqrt{k} \right|, \tag{2.75}$$

[3] See exercise 1 on p. 149 for an explanation of this term and for more details.

where l, j are integers. Now, we use our assumption that $\sqrt{k}$ is some noninteger rational number p/q:

$$D_N = \left| l + j\frac{p}{q} \right|, \tag{2.76}$$

where p, q are integers. According to the lemma, we must have $D_N \geq 1/q$ for any values of l, j (which implies any value of N). However, this contradicts our previous conclusion that D_N becomes arbitrarily small for large values of N.

Since rational and irrational numbers obey the same rules for all operations, most mathematical problems are solved without even thinking about the types of numbers they use. We solve a quadratic equation $ax^2 + bx + c = 0$ the same way, whether its coefficients a, b, and c are rational or irrational numbers.

Applications for real numbers are difficult to list because they are countless (pun intended). They originate in our ability and desire to *measure* things, whether it is the output of a factory, the temperature outside, or the intensity of cosmic rays.

By applying mathematical models to measured data, we can make *predictions*. We can forecast weather, predict the time of sunrise, or the existence of antimatter, as Paul Dirac did in 1928. Today, much of the new technology is first designed and tested by using mathematical models, which are cheaper and quicker to make than a hardware prototype. Predictions made by mathematical models have become vital to our well-being.

Another great ability of math is to *infer* things. From the amount of carbon-14 in a small specimen of wood in an ancient burial, we can infer its age, and from the number of web searches for a cold medicine, we can infer the spread of flu.

Why do mathematics in general and numbers in particular have this magic power? Maybe the answer is given by Galileo, who famously wrote that the Book of Nature is written in the language of mathematics. At his time, this meant mostly geometry. Today, with our knowledge of modern mathematics, the words of Galileo ring even more true.

2.6 Two Theorems about Real Numbers

Theorem 2.19 *Any two numbers x and y can be uniquely represented as a sum and a difference of some other two numbers u and v:*

$$\begin{aligned} x &= u + v, \\ y &= u - v. \end{aligned} \tag{2.77}$$

Proof
This is a constructive proof. We define numbers u and v as follows:

$$u = \frac{x+y}{2}, \quad v = \frac{x-y}{2}. \tag{2.78}$$

Then

$$u+v = \frac{x+y}{2} + \frac{x-y}{2} = x, \quad u-v = \frac{x+y}{2} - \frac{x-y}{2} = y. \tag{2.79}$$

Numbers u and v are constructed in a way that guarantees their uniqueness. Indeed, if we add and subtract equations (2.79) we will get the original definitions of u and v.

Theorem 2.20 *Any two positive numbers x and y can be represented as a product and a ratio of some other two numbers p and q:*

$$x = pq, \quad y = \frac{p}{q}. \tag{2.80}$$

Proof
The proof is analogous to that of the previous theorem. We construct numbers p and q as follows: $p = \sqrt{xy}$ *and* $q = \sqrt{x/y}$.

$$pq = \sqrt{xy}\sqrt{\frac{x}{y}} = x, \quad \frac{p}{q} = \frac{\sqrt{xy}}{\sqrt{\frac{x}{y}}} = y. \tag{2.81}$$

Note that though this theorem is similar to the previous one, it does not claim uniqueness of p and q.

2.7 Complex Numbers

The need for the next extension of the set of numbers arises when we try to solve an equation like $x^2 = a$ for all real values of a. If $a \geq 0$, there are two solutions that are given by $\pm\sqrt{a}$. However, if $a < 0$, there are no real numbers that satisfy this equation. We define i as such a number that $i^2 = -1$. This number is called the *imaginary unit.* We also stipulate that all standard properties of addition and multiplication are valid for i alone or in combination with real numbers. Therefore, we can multiply i by itself, or by a real number, add imaginary and real numbers, and so on. These operations will be commutative, associative, and distributive, just as in the case of real numbers. An expression in the form $x + iy$, where x and y are real, is called a complex number. In this complex number, x is called its real part, and y is called its imaginary part. They are denoted as $x = \mathrm{Re}(z); y = \mathrm{Im}(z)$.

> The adjective "imaginary" does a disservice to these numbers. They are as real as their "real" counterparts. The laws of quantum mechanics for the behavior of atoms and molecules are described by the Schrödinger equation (or its relativistic cousin, the Dirac equation), prominently featuring the imaginary unit i. Applications of complex numbers in technology are also common. Almost any radio wave signal, such as one used by cell phones, is described by oscillating trigonometric functions, sine and cosine. Because of Euler's formula (see section 7.17), sines and cosines can be represented through a complex exponent, which is much easier to work with than trigonometric functions. That's why signal processing engineers commonly talk about the "real" and "imaginary" components of a radio signal that is, in fact, a quite real electromagnetic wave.

A sum of two complex numbers $z_1 = x_1 + iy_1$ and $z_2 = x_2 + iy_2$ is also a complex number that is defined as

$$z_1 + z_2 = (x_1 + x_2) + i(y_1 + y_2). \tag{2.82}$$

We also define a product of two complex numbers as

$$z_1 z_2 = (x_1 x_2 - y_1 y_2) + i(x_1 y_2 + x_2 y_1). \tag{2.83}$$

This way of computing the sum and product of complex numbers is "backward compatible" with the summation and multiplication of real numbers. This means that the result of adding or multiplying two real numbers does not depend on whether we do it directly or view each real number as a complex number with a zero imaginary part. Indeed, from equation (2.82) we get for the sum of two real numbers:

$$\begin{aligned}(x_1 + i \cdot 0) + (x_2 + i \cdot 0) &= (x_1 + x_2) + i \cdot (0 + 0) \\ &= x_1 + x_2,\end{aligned} \tag{2.84}$$

which is the same as a sum of two real numbers that is computed directly. Similarly, we get from equation (2.83) the following result for the product of two real numbers that are treated as complex numbers with the zero imaginary part:

$$\begin{aligned}(x_1 + i \cdot 0)(x_2 + i \cdot 0) &= (x_1 x_2 - 0 \cdot 0) + i(x_1 \cdot 0 + x_2 \cdot 0) \\ &= x_1 x_2.\end{aligned} \tag{2.85}$$

Again, we get the expected result x_1x_2. We will return to multiplication of complex numbers in section 7.15 and will show there that it is also compatible with the fact that the product of two negative real numbers is positive.

In addition to "backward compatibility," complex numbers possess all the major properties of real numbers. This is proved by the following theorems.

Theorem 2.21 *Addition and multiplication of complex numbers are commutative.*

Proof
For a sum of two complex numbers we have:

$$(a+ib)+(c+id)=(a+c)+i(b+d). \tag{2.86}$$

It is easy to see that we get the same final result for $(c+id)+(a+ib)$. For a product we have

$$(a+ib)(c+id)=(ac-bd)+i(ad+bc). \tag{2.87}$$

Again, we get the same final result for $(c+id)(a+ib)$.

Theorem 2.22 *(Distributivity) For any three complex numbers z_1, z_2, and z_3 it is the case that*

$$z_1(z_2+z_3)=z_1z_2+z_1z_3. \tag{2.88}$$

Proof
We consider $z_j=x_j+iy_j$, where $j=1,2,3$. On one hand, we have

$$\begin{aligned}
z_1(z_2+z_3) &= (x_1+iy_1)((x_2+iy_2)+(x_3+iy_3))\\
&= (x_1+iy_1)((x_2+x_3)+i(y_2+y_3))\\
&= (x_1(x_2+x_3)-y_1(y_2+y_3))+i(y_1(x_2+x_3)+x_1(y_2+y_3))\\
&= (x_1x_2+x_1x_3-y_1y_2-y_1y_3)+i(y_1x_2+y_1x_3+x_1y_2+x_1y_3).
\end{aligned} \tag{2.89}$$

On the other hand we have

$$\begin{aligned}
z_1z_2+z_1z_3 &= (x_1+iy_1)(x_2+iy_2)+(x_1+iy_1)(x_3+iy_3)\\
&= (x_1x_2-y_1y_2)+i(x_1y_2+y_1x_2)+(x_1x_3-y_1y_3)+i(x_1y_3+y_1x_3)\\
&= (x_1x_2+x_1x_3-y_1y_2-y_1y_3)+i(y_1x_2+y_1x_3+x_1y_2+x_1y_3).
\end{aligned} \tag{2.90}$$

We compare the right-hand sides of equations (2.89) and (2.90) and see that they are identical. This proves the distributive property.

Theorem 2.23 *(Associativity.) For any three complex numbers z_1, z_2, and z_3 it is the case that*

$$\begin{aligned}
z_1(z_2z_3) &= (z_1z_2)z_3,\\
z_1+(z_2+z_3) &= (z_1+z_2)+z_3.
\end{aligned} \tag{2.91}$$

Proof

The proof of associativity for addition is straightforward. We have

$$\begin{aligned}(x_1+iy_1)+((x_2+iy_2)+(x_3+iy_3)) &= (x_1+iy_1)+(x_2+x_3)+i(y_2+y_3)\\ &= (x_1+x_2+x_3)+i(y_1+y_2+y_3).\end{aligned} \tag{2.92}$$

On the other hand, we have

$$\begin{aligned}((x_1+iy_1)+(x_2+iy_2))+(x_3+iy_3) &= (x_1+x_2)+i(y_1+y_2)+(x_3+iy_3)\\ &= (x_1+x_2+x_3)+i(y_1+y_2+y_3).\end{aligned} \tag{2.93}$$

We compare the right-hand sides of the two last equations and see that they are identical, which proves associativity for addition.

Proving associativity for multiplication is a bit more cumbersome. Again, we consider the two expressions separately. On one hand,

$$\begin{aligned}((x_1+iy_1)(x_2+iy_2))(x_3+iy_3) =&((x_1x_2-y_1y_2)+i(x_1y_2+y_1x_2))(x_3+iy_3)\\ =&((x_1x_2-y_1y_2)x_3-(x_1y_2+y_1x_2)y_3)\\ &+i((x_1x_2-y_1y_2)y_3+(x_1y_2+y_1x_2)x_3)\\ =&(x_1x_2x_3-y_1y_2x_3-x_1y_2y_3-y_1x_2y_3)\\ &+i(x_1x_2y_3-y_1y_2y_3+x_1y_2x_3+y_1x_2x_3).\end{aligned} \tag{2.94}$$

On the other hand, we have

$$\begin{aligned}(x_1+iy_1)((x_2+iy_2)(x_3+iy_3)) =&(x_1+iy_1)((x_2x_3-y_2y_3)+i(x_2y_3+y_2x_3))\\ =&(x_1(x_2x_3-y_2y_3)-y_1(x_2y_3+y_2x_3))\\ &+i(y_1(x_2x_3-y_2y_3)+x_1(x_2y_3+y_2x_3))\\ =&(x_1x_2x_3-y_1y_2x_3-x_1y_2y_3-y_1x_2y_3)\\ &+i(x_1x_2y_3-y_1y_2y_3+x_1y_2x_3+y_1x_2x_3).\end{aligned} \tag{2.95}$$

Again, we see that the right-hand sides of the two last equations are identical, which proves associativity for multiplication.

Because of these theorems, it is not necessary to memorize formula (2.83) for the product of two complex numbers. We can just distribute the factors and replace i^2 with -1 in the result:

$$\begin{aligned}(a+ib)(c+id) &= ac+ibc+iad+i^2bd\\ &= (ac-bd)+i(bc+ad).\end{aligned} \tag{2.96}$$

We get a complex number with the real part $(ac-bd)$ and the complex part $(bc+ad)$.

It is common to illustrate a complex number as a point on a plane with the horizontal coordinate given by the real part, and the vertical coordinate given by the imaginary part. Looking at this graphical representation, we can ask how far this point is located from the origin. The distance from the point (x, y) to the origin is given by the Pythagoras theorem: $\sqrt{x^2 + y^2}$. For a complex number $z = x + iy$, the real-valued nonnegative value given by $\sqrt{x^2 + y^2}$ is called the modulus: $|z| = \sqrt{x^2 + y^2}$.

> The first person to show a complex number as a point on a plane was Jean-Robert Argand. He was not a professional mathematician, but managed a bookshop in Paris in the beginning of the 19th century.

One useful property of complex numbers deals with *complex conjugate* numbers. For a complex number $z = x + iy$, the complex conjugate is defined as $z^* = x - iy$. If we use equation (2.83) to compute the product of a number $z = x + iy$ and its conjugate, we will always get a real number—that is, the imaginary part of the product will be equal to zero:

$$\begin{aligned} z^* \cdot z &= (x^2 + y^2) + i(xy - xy) \\ &= x^2 + y^2. \end{aligned} \tag{2.97}$$

Moreover, we see that $z^* z = |z|^2$. This property is used to explicitly determine the real and imaginary parts for the ratio of complex numbers:

$$\begin{aligned} \frac{z_1}{z_2} &= \frac{z_1 z_2^*}{z_2 z_2^*} \\ &= \frac{z_1 z_2^*}{|z_2|^2} \\ &= \frac{x_1 x_2 + y_1 y_2}{|z_2|^2} + i \frac{y_1 x_2 - x_1 y_2}{|z_2|^2}. \end{aligned} \tag{2.98}$$

Using complex conjugates, we can compute the real and imaginary parts of a complex number as

$$\begin{aligned} \text{Re}(z) &= \frac{z + z^*}{2}, \\ \text{Im}(z) &= \frac{z - z^*}{2}. \end{aligned} \tag{2.99}$$

Theorem 2.24 *For any two complex numbers z_1 and z_2 it is the case that*

$$\begin{aligned} (z_1 z_2)^* &= z_1^* z_2^*, \\ (z_1 + z_2)^* &= z_1^* + z_2^*, \\ (z^*)^* &= z. \end{aligned} \tag{2.100}$$

A proof of this theorem is a subject of exercise 18 for this chapter.

Below we prove three theorems for the modulus:

Theorem 2.25 *For any two complex numbers z_1, z_2 it is the case that*

$$|z_1||z_2| = |z_1 z_2|. \tag{2.101}$$

Proof
Consider two complex numbers

$$\begin{aligned} z_1 &= x_1 + iy_1, \\ z_2 &= x_2 + iy_2. \end{aligned} \tag{2.102}$$

Their product is given by

$$z_1 z_2 = (x_1 x_2 - y_1 y_2) + i(x_1 y_2 + x_2 y_1). \tag{2.103}$$

The modulus of the product is

$$\begin{aligned} |z_1 z_2| &= \sqrt{(x_1 x_2 - y_1 y_2)^2 + (x_1 y_2 + x_2 y_1)^2} \\ &= \sqrt{x_1^2 x_2^2 + y_1^2 y_2^2 - 2x_1 x_2 y_1 y_2 + x_1^2 y_2^2 + x_2^2 y_1^2 + 2x_1 y_2 x_2 y_1} \\ &= \sqrt{x_1^2 x_2^2 + y_1^2 y_2^2 + x_1^2 y_2^2 + x_2^2 y_1^2}. \end{aligned} \tag{2.104}$$

Separately, the product of the moduli $|z_1|$ and $|z_2|$ is computed as follows:

$$\begin{aligned} |z_1||z_2| &= \sqrt{(x_1^2 + y_1^2)(x_2^2 + y_2^2)} \\ &= \sqrt{x_1^2 x_2^2 + x_1^2 y_2^2 + x_2^2 y_1^2 + y_1^2 y_2^2}. \end{aligned} \tag{2.105}$$

Comparison of equations (2.104) and (2.105) proves the theorem.

Another proof is based on the Brahmagupta–Fibonacci identity and is the subject of exercise 1 for chapter 5. In section 7.17 we will also show a connection between this theorem and Euler's formula.

Theorem 2.26 *If $z_1 z_2 = 0$, then $z_1 = 0$ or $z_2 = 0$.*

Proof
This theorem directly follows from theorem 2.25.

Theorem 2.27 *For any two complex numbers z_1, z_2, it is the case that*[4]

$$|z_1 + z_2| \leq |z_1| + |z_2|. \tag{2.106}$$

[4] Here we rely on several familiar properties of inequalities. We will consider inequalities in more detail in chapter 3.

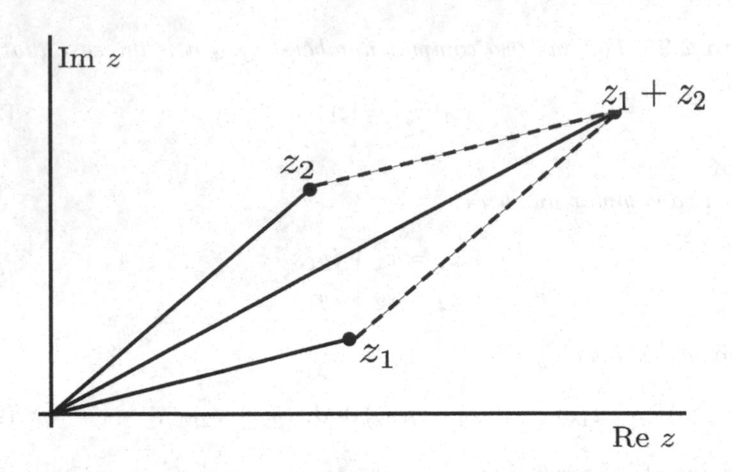

FIGURE 2.2
A graphic representation for a sum of complex numbers

We give three proofs of this theorem. The first proof is graphical:

Proof
This theorem is evident from figure 2.2. Numbers z_1 and z_2 form two sides of a triangle, and number $z_1 + z_2$ forms the third side. We know that the sum of the lengths of two sides of a triangle is always greater or equal than the length of the third side, which proves the theorem.

The second proof is algebraic.

Proof
We need to prove the following inequality:

$$\sqrt{(x_1 + x_2)^2 + (y_1 + y_2)^2} \leq \sqrt{x_1^2 + y_1^2} + \sqrt{x_2^2 + y_2^2}. \tag{2.107}$$

Both sides are nonnegative, so we can square this inequality and preserve its sign:

$$(x_1 + x_2)^2 + (y_1 + y_2)^2 \leq x_1^2 + y_1^2 + x_2^2 + y_2^2 + 2\sqrt{(x_2^2 + y_2^2)(x_1^2 + y_1^2)}. \tag{2.108}$$

We expand the squares in the left-hand side and cancel common terms in the left-hand and right-hand sides:

$$2x_1x_2 + 2y_1y_2 \leq 2\sqrt{(x_2^2 + y_2^2)(x_1^2 + y_1^2)}. \tag{2.109}$$

Next, we divide this inequality by 2 and square it again to get rid of the square root:

$$x_1^2x_2^2 + y_1^2y_2^2 + 2x_1x_2y_1y_2 \leq (x_2^2 + y_2^2)(x_1^2 + y_1^2). \tag{2.110}$$

We expand the parentheses in the right-hand side. This yields

$$x_1^2x_2^2 + y_1^2y_2^2 + 2x_1x_2y_1y_2 \leq x_1^2x_2^2 + y_1^2y_2^2 + x_1^2y_2^2 + y_1^2x_2^2. \tag{2.111}$$

Common terms $x_1^2x_2^2+y_1^2y_2^2$ *cancel. We move all remaining terms to the right-hand side:*

$$0 \leq x_1^2y_2^2 + y_1^2x_2^2 - 2x_1x_2y_1y_2. \tag{2.112}$$

In the right-hand side, we see a complete square:

$$0 \leq (x_1y_2 - y_1x_2)^2. \tag{2.113}$$

It is nonnegative, which proves the theorem.[5]

Proof
We start the third proof by observing that the modulus of a complex number is greater or equal than the absolute value of each of its components—that is, for $z = x + iy$ *we have*

$$\begin{aligned} |z| &\geq |\mathrm{Re}(x)|, \\ |z| &\geq |\mathrm{Im}(x)|. \end{aligned} \tag{2.114}$$

Next, we use complex conjugates to compute the square of the modulus of a sum of two complex numbers (note that we use equation (2.97) here):

$$|z_1 + z_2|^2 = (z_1 + z_2)(z_1^* + z_2^*). \tag{2.115}$$

We expand the right-hand side by the distributive property:

$$|z_1 + z_2|^2 = z_1z_1^* + z_2z_1^* + z_2^*z_1 + z_2z_2^*. \tag{2.116}$$

This is the same as (see equation (2.97)).

$$|z_1 + z_2|^2 = |z_1|^2 + z_2z_1^* + (z_2z_1^*)^* + |z_2|^2. \tag{2.117}$$

We use the first equation in (2.99) to transform the second and third term in the right-hand side of equation (2.117):

$$|z_1 + z_2|^2 = |z_1|^2 + 2\mathrm{Re}(z_2z_1^*) + |z_2|^2. \tag{2.118}$$

Then inequality (2.114) produces:

$$|z_1 + z_2|^2 \leq |z_1|^2 + 2|z_2z_1^*| + |z_2|^2. \tag{2.119}$$

---

[5]Note a similarity with the proof of the Cauchy-Schwarz inequality for two pairs of numbers on p. 75.

Theorem 2.25 yields $|z_2 z_1^*| = |z_2||z_1^*| = |z_2||z_1|$. *Then*

$$|z_1 + z_2|^2 \leq |z_1|^2 + 2|z_2||z_1| + |z_2|^2. \tag{2.120}$$

The right-hand side is a complete square:

$$|z_1 + z_2|^2 \leq (|z_1| + |z_2|)^2. \tag{2.121}$$

We extract the square root from both sides of this inequality to get the statement of the theorem.

So far, we have been successful in extending the set of numbers from natural to integers, then to rationals, to reals, and finally to complex. This prompts a question whether we can extend this set further and come up with some new powerful super-complex numbers, while keeping all the properties of addition and multiplication (commutativity, associativity, and distributivity). In section 2.8.7 we will learn that real numbers do not have any gaps or holes (unlike rational numbers). This means that we cannot "squeeze in" more numbers into the number line. Complex numbers are defined by adding another axis—that is, extending the realm of numbers to two dimensions. Maybe we can define a number that is represented as a point in a 3D space? Interestingly, the answer is no. We can define an object that is represented as a point in a 3D or 4D space, but then some of the nice properties of addition and multiplication must be sacrificed. We will talk about some reasons for this later (p. 171). This shows that mathematical objects have their intrinsic properties, and we cannot assign desired properties to them at will. It looks like our job is to *discover* math, but our ability to *create* math may be limited by math itself.

For mathematical objects that do not possess all the useful properties of real and complex numbers, such as matrices,[6] rules of computation become cumbersome. Consider, for example, what happens if the multiplication of two matrices $\mathcal{A}$ and $\mathcal{B}$ is no longer commutative. Then the familiar formula for the square of a sum looks different:

$$\begin{aligned}(\mathcal{A} + \mathcal{B})^2 &= (\mathcal{A} + \mathcal{B})(\mathcal{A} + \mathcal{B}) \\ &= \mathcal{A}\mathcal{A} + \mathcal{A}\mathcal{B} + \mathcal{B}\mathcal{A} + \mathcal{B}\mathcal{B} \\ &= \mathcal{A}^2 + \mathcal{A}\mathcal{B} + \mathcal{B}\mathcal{A} + \mathcal{B}^2,\end{aligned} \tag{2.122}$$

where we no longer are able to simplify this expression because $\mathcal{A}\mathcal{B} + \mathcal{B}\mathcal{A} \neq 2\mathcal{A}\mathcal{B}$. If we square the sum of three or more matrices, the result becomes even more cumbersome.

From now on we will assume that all numbers we use in this book are real, unless we explicitly state that they can be complex or that they are natural, integer, or rational.

[6] Matrix is a set of numbers arranged in a rectangular array. Algebra defines rules for adding and multiplying matrices.

2.8 How Many Numbers Are There?

2.8.1 Sets

The notion of a *set* is one of the most fundamental and important in mathematics. A set is a well-defined collection of distinct objects, considered as an object in its own right. For example, the numbers 2, 4, and 6 are distinct objects when considered separately, but when they are considered collectively they form a single set of size three, written as {2, 4, 6}.

The elements of a set are not necessarily numbers. The White House, the King of England, and the star Aldebaran form a set. A set can contain other set(s) as its elements. A set can have a finite or infinite number of elements. The important condition is that we should be able to determine if any particular object is an element of that set or not.[7] For example, a set of integer numbers is an infinite set that includes number 11780, but does not include number 0.5 or the star Aldebaran.

There is only one set that does not contain any elements. It is called the empty set. There cannot be two empty sets because there would be no way to discriminate one empty set from another.[8]

We say that two sets have the same *cardinality* if we can define a mapping between these two sets that is one-to-one in both directions. This means that for each element of the first set we have a corresponding one (and only one) element of the second set, and for each element of the second set we have a corresponding one (and only one) element of the first set. Such a mapping is called a *bijection*. For example, sets {21, 14, 66} and {the White House, the King of England, and the star Aldebaran} have the same cardinality because we can define a bijection between them:

$$\begin{aligned} 21 &\leftrightarrow \text{the White House,} \\ 14 &\leftrightarrow \text{the King of England,} \\ 65 &\leftrightarrow \text{Aldebaran.} \end{aligned}$$

Compare this with the case when the first set has four elements:

$$\begin{aligned} 21 &\leftrightarrow \text{the White House,} \\ 14 &\leftrightarrow \text{the King of England,} \\ 33 &\searrow \\ 65 &\leftrightarrow \text{Aldebaran.} \end{aligned}$$

[7] This is not the only condition to define a set, but here we will not go into most intricate details.

[8] The uniqueness of the empty set gives mathematicians a great way to define the number 0. From this starting point, one may construct other sets that correspond to numbers 1, 2, 3, and so on.

In this case for each element of the set $\{21, 14, 33, 65\}$ there is one element of the set {the White House, the King of England, and the star Aldebaran}, but not the other way around. In plain English, we may say that if two sets have the same cardinality, they have the same number of elements. However, cardinality is a more general concept because it permits a comparison of infinite sets as well as finite sets.

Sets that have the same cardinality as the set of natural numbers, are called *countable*. For any two countable sets, we can establish a bijection between them. Such a mapping, while mathematically valid, can produce counterintuitive conclusions. Consider the set of natural numbers and the set of even natural numbers. The latter is a subset of the former, and we expect it to have "fewer" elements. Yet, both sets are countable because we can enumerate the set of even numbers:

$$
\begin{gathered}
1 \leftrightarrow 2 \\
2 \leftrightarrow 4 \\
3 \leftrightarrow 6 \\
\ldots \\
n \leftrightarrow 2n \\
\ldots
\end{gathered}
$$

David Hilbert illustrated the strange world of countability in his Grand Hotel paradox. Imagine a hotel that has infinitely many numbered rooms. On a particular day, all the rooms are occupied, but a new guest arrives at the reception. The manager of the hotel tells the guest that there are no free rooms available, but she can still accommodate the new arrival by moving the current guests a bit. The guest in room 1 is asked to move to room 2, the guest in room 2 is asked to move to room 3, and so on. This frees room 1 for the new guest.

On the next day, an infinite number of guests arrive at the hotel. The manager still tells them that there is no problem: a current guest in room 1 is asked to move to room 2, the guest in room 2 is asked to move to room 4, the guest in room 3 is asked to move to room 6, and so on. This move frees all odd-numbered rooms, where the new guests can stay. (The logistics of such a move was not a concern for David Hilbert.)

2.8.2 How Many Rational Numbers Are There?

From section 2.8.1, we can see that there is a set of integers and there is a set of rational numbers. Indeed, for any number we can determine whether it belongs to the set of integers, to the set of rationals, or does not belong to these sets. Now, we can ask a question: which set has a larger cardinality? (In plain English: are there more rational numbers than there are integers?)

We know that rational numbers can be arbitrarily close to each other. If we take any rational number, say, 2.1249876987, we can construct another rational number that differs from the first one by a very small amount, such as 10^{-6} or 10^{-100} or $10^{-10^{10}}$. There are infinitely many rational numbers between 0 and 1, or between any other two different numbers.

Intuitively, it appears that rational numbers must have higher cardinality. First, any integer is a rational number, so integers form a subset of rational numbers. Second, between any two consecutive integers there are infinitely many rational numbers. This is a notable example of when our intuition fails. The cardinality of these two sets is the same.

Theorem 2.28 *The rational numbers form a countable set.*

Proof

By definition, a rational number can be represented as m/n, where m, n are integers. Let's put all rational numbers in a 2D lattice, where the horizontal axis will correspond to m and the vertical axis to n. Each rational number will correspond to a point on that lattice (figure 2.3). For example, rational number 0 will be placed at $(m = 0, n = 1)$, rational number 1 is placed at $(m = 1, n = 1)$ and so on. Note that this lattice contains all rational numbers; some of them are shown in the figure.

Next, we number these points in a spiral, as shown in the figure using arrows. Each rational number will correspond to more than one point on the lattice. For example the point $(m = 1, n = 1)$ corresponds to the same rational number 1 as the point $(m = 2, n = 2)$. When numbering points, we only use the first occurrence of a particular rational number; these points are shown with filled circles. As we go in the spiral, we skip all points with $n = 0$ or those that we have numbered before. The circles for skipped points are not filled in the figure. Points on the lattice that have $n = 0$ do not correspond to any rational numbers; they are also shown with empty circles.

It is clear that this way we can enumerate all rational numbers—that is, for any rational number we have a corresponding natural number, and for any natural number we have a corresponding rational number. This procedure establishes a bijection between rational numbers and natural numbers.

Rational numbers have the same cardinality as natural numbers!

2.8.3 How Many Real Numbers Are There?

If rational numbers unexpectedly form a countable set, what can we say about real numbers? Are they countable as well? The next theorem gives the answer to that question. Its beautiful proof by Georg Cantor is called "the diagonal argument."

Theorem 2.29 *The set of real numbers is not countable.*

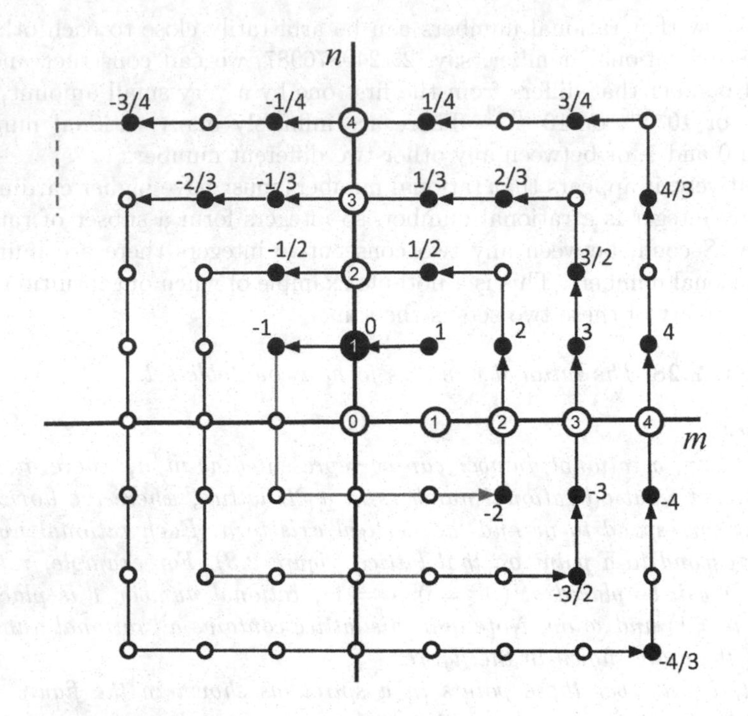

FIGURE 2.3
Countability of rational numbers

Proof

The proof is by contradiction. We consider an open interval between zero and one—that is, all numbers that are greater than 0 *and smaller than* 1. *Contrary to the statement of the theorem, we assume that real numbers in this interval are countable. Then they can be arranged in an infinite list, starting from the first row and down. Indeed, the countability means that we can define the first real number, the second one, and so on.*

We will use a decimal representation of real numbers with an infinite number of digits. If a number has a finite decimal representation, we will augment it with zeros. Here is how such a list looks:

$$\begin{aligned} &0.a_1a_2a_3a_4a_5a_6\cdots \\ &0.b_1b_2b_3b_4b_5b_6\cdots \\ &0.c_1c_2c_3c_4c_5c_6\cdots \\ &0.d_1d_2d_3d_4d_5d_6\cdots \\ &\quad\vdots \end{aligned} \tag{2.123}$$

where $a_1a_2a_3a_4a_5a_6$ *are the digits in the first real number,* $b_1b_2b_3b_4b_5b_6$ *are the digits in the second real number and so on. All of these digits are from*

0 *to* 9. *For example, if the first number is* 0.76573, *the digits would be:* $a_1 = 7; a_2 = 6; a_3 = 5; a_4 = 7; a_5 = 3; a_6 = 0$, *and all subsequent digits* $(a_7, a_8, \cdots)$ *are zeroes. Next, we define a change to any digit. Such a change can be done in many different ways, for example, we can stipulate that any digit that is not equal to* 5, *we change to* 5, *and if a digit is equal to* 5, *we change it to* 6. *Here the important point is that a digit does not remain the same. Finally, we construct one particular real number* X. *We take the first digit from the first row, change it according to the rules above and use it as the first digit in number* X. *Then, we take the second digit from the second real number, apply the change and use it as the second digit in* X, *and so on. We will get the following number:*

$$X = 0.a_1'b_2'c_3'd_4'\cdots \tag{2.124}$$

where a prime superscript means that a digit was changed.

Will the number X *be in the list? Can it be the number in the first row of the list? No, because its first digit does not match* a_1. *Can it be the number in the second row? Still no, because its second digit does not match* b_2. *Can it be the number in row number* 2875*? No, because digit number* 2875 *does not match. We must therefore conclude that* X *is not in the list. However, this contradicts our initial assumption that all real numbers from* 0 *to* 1 *have been arranged in this list. This contradiction means that our initial assumption was wrong, and we must conclude that the set of real numbers in the interval* 0 *to* 1 *is not countable.*

We established that real numbers from 0 to 1 are not countable. It is clear that this implies that all real numbers (ranging from minus infinity to plus infinity) are also not countable. In plain English, there are "more" real numbers than rational or integer ones.

2.8.4 Cardinality of Real Numbers in an Interval

Here, we consider open and closed intervals in real numbers. The open interval is all real numbers between a and b, not including these end values. (An actual definition is more complex, but we will not use it.) A half-closed interval includes one of the end points. A closed interval includes both end points.

Real numbers range from $-\infty$ to $+\infty$. They, of course, form a set, which is denoted as $\mathbb{R}$. We can also define a set of real numbers in any particular interval, say, from -1 to 1 (but not including these values); let's denote it $I_{-1,1}$.

Theorem 2.30 *Sets* $\mathbb{R}$ *and* $I_{-1,1}$ *have the same cardinality.*

Proof

To prove that, we need to define a bijection mapping between these sets. One

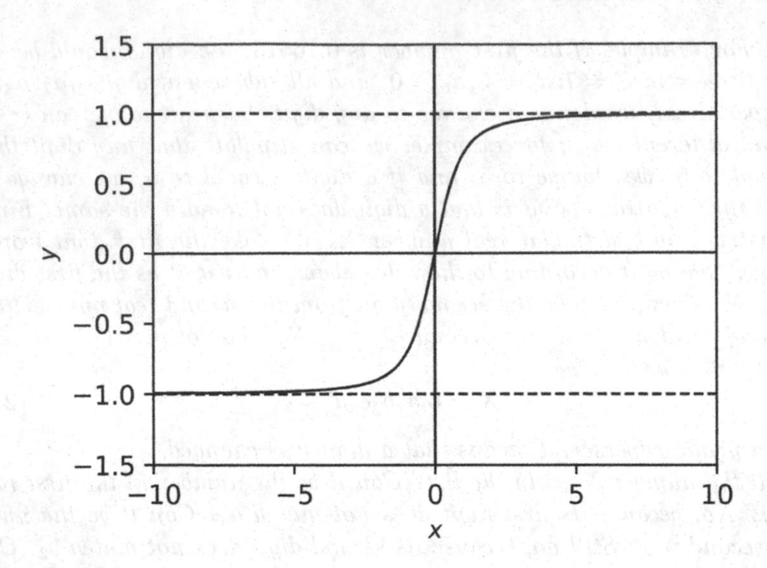

FIGURE 2.4
Bijection between all real numbers and the interval $R_{-1,1}$

example of such mapping is:

$$\begin{aligned} &x \in \mathbb{R}, \textit{ (reads: } x \textit{ is in set } \mathbb{R}), \\ &y \in I_{-1,1}, \textit{ (reads: } y \textit{ is in set } I_{-1,1}), \\ &y = \frac{x}{\sqrt{x^2+1}}. \end{aligned} \tag{2.125}$$

Equations (2.125) define a mapping from x to y, but we can solve the last equation in (2.125) to produce a mapping from y to x:

$$x = \frac{y}{\sqrt{1-y^2}}, \tag{2.126}$$

Equations (2.125) and (2.126) create a bijection, where x ranges from $-\infty$ to $+\infty$, and y ranges from -1 to 1 (not including the end points). This mapping is shown in figure 2.4.

In a similar way, we can map the whole set of real numbers to any open interval, no matter how small! We modify the last equation in (2.125) as follows:

$$y = a + \frac{bx}{\sqrt{x^2+1}}. \tag{2.127}$$

Equation (2.127) also has an inverse, which guarantees having a bijection, but the value of y now ranges from $a-b$ to $a+b$, where b can be arbitrarily small. This way the entire number line can be "squeezed into" an interval that ranges, for example, from -10^{-100} to $+10^{-100}$.

Next we consider a closed and half-closed interval. A way to map a closed interval R of real numbers to a half-closed interval R_1 is given by the following theorem:

Theorem 2.31 *A closed and a half-closed interval of real numbers have the same cardinality.*

Proof

We consider a half-closed interval I from 0 to 1—that is, all numbers y between 0 and 1, but not including point $y = 1$. We map it to a closed interval I_1, which has all points from 0 to 1, including the end points. The two intervals differ by one end point, which is included in the closed interval, but not in the half-closed one. Basically, our goal is to "free up" one real number, and map the rest. The "freed up" value can be used to map that end point.

We get a cue from how the manager of Hilbert's Grand Hotel "freed up" a room. For natural numbers this is done by mapping two sets with a shift:

$$\begin{aligned} 0 &\leftrightarrow 1 \\ 1 &\leftrightarrow 2 \\ 2 &\leftrightarrow 3 \\ &\dots \end{aligned} \tag{2.128}$$

Within the closed interval, we select a countable subset of values $x_n = 1/2^n$, where $n = 0, 1, 2, \ldots$. All these points are located within the closed interval.[9] *Within the half-closed interval, we select the same points $y_n = 1/2^n$, but $n = 0$ is excluded. We map these points with a shift:*

$$\begin{aligned} x_0 &\leftrightarrow y_1 \\ x_1 &\leftrightarrow y_2 \\ x_2 &\leftrightarrow y_3 \\ &\dots \end{aligned} \tag{2.129}$$

This process is shown in figure 2.5 using solid line arrows.

All points that are not equal to $1/2^n$ are mapped as $y = x$. This is shown with dashed line arrows in the figure. This completes the mapping of a closed interval to a half-closed one.

2.8.5 Points on a Plane

A point on a plane is a pair of two real numbers that is given by coordinates x and y. Such pairs are *ordered*, which means that pair (x, y) is not the same as pair (y, x). There is a set of such pairs, called $\mathbb{R}^2$. How many of those are there? Is the cardinality of this set the same as the cardinality of $\mathbb{R}$ or higher?

[9] As we shall see in section 6.1, $1/2^0 = 1$, so this value belongs to the closed interval.

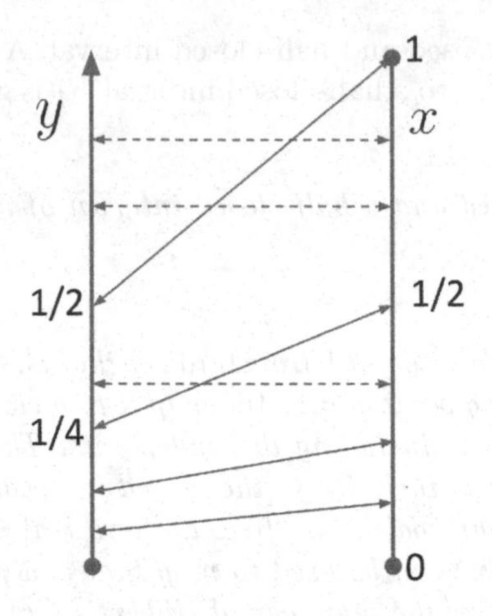

FIGURE 2.5
Mapping of a closed interval to a semi-closed one

Theorem 2.32 *The cardinality of the set of ordered pairs of real numbers is the same as the cardinality of the set of real numbers.*

Proof
Here is a "less than rigorous" proof. We will build a correspondence between the two sets: an open interval $(0, 1)$ *and a pair of such intervals. Let's assume that we have number* x *in the first set:* $x \in I_{0,1}$. *We also have* (y, z) *in the second set:* $(y, z) \in I_{0,1}^2$. *How do we map* x *into* (y, z)*? To do that, we use decimal representations for* (y, z)*:*

$$\begin{aligned} y =& 0.b_1 b_2 b_3 b_4 b_5 b_6 \cdots \\ z =& 0.c_1 c_2 c_3 c_4 c_5 c_6 \cdots \end{aligned} \tag{2.130}$$

Next, we map this into a one-dimensional interval by interleaving digits. We take a digit from y *and a digit from* z *and then alternate them to get a decimal representation for* x*:*

$$x = 0.b_1 c_1 b_2 c_2 b_3 c_3 b_4 c_4 \cdots \tag{2.131}$$

If we need to map x *back to* (y, z)*, we select odd digits for* y *and even digits for* z*. This almost accomplishes our goal to construct a bijection between a 1D and a 2D intervals for real numbers and almost proves that these sets have the same cardinality. Unfortunately, there still are points, which are not mapped correctly. For example, consider numbers* $0.0091919191\ldots$ *and* $0.1001010101\ldots$*. If we try to split them into pairs, we will get* $0.0999999\ldots$ *and* $0.01111\ldots$ *for the first number, and* $0.100000\ldots$ *and* $0.01111\ldots$ *for the second one. However, from equation (2.43) it follows that these two pairs are identical. This means that the correspondence between real numbers in the interval and pairs of number is not truly one-to-one. This flaw is fixable, but the details of this fix are beyond the scope of this book.*

2.8.6 Numbers That We Can Define

We already know that rational numbers form a countable set. In plain English, there are "as many" rational numbers as there are integers.

What about adding to that set some irrational numbers that we can define? We can define numbers $\sqrt{2}$ or $\sqrt[5]{876876}$. (Such numbers are solutions of polynomial equations with integer coefficients and are called algebraic numbers.) We can also define a number using some kind of algorithm or procedure. Let's suppose that any procedure, as written in a document, must have a finite length and must use a fixed and finite alphabet in order to be documented. We may call a number definable if it can be computed to arbitrary precision using some kind of recipe.

For example, the number 2 is defined by character "2." This is a finite procedure to define a number with absolute precision. In another example, let's define the following procedure:

1. Start from a positive number, for example $s_1 = 2$.
2. Compute the next number by plugging the current one into the right-hand side of the following equation:

$$s_2 = \frac{1}{2}\left(s_1 + \frac{X}{s_1}\right). \tag{2.132}$$

3. Continue these iterations, getting new numbers.

Turns out, this procedure produces ever better approximations for $\sqrt{X}$. It would require an infinite number of steps to get infinite accuracy, but the *recipe* for it is finite and given above. Also, a computation to arbitrarily high finite accuracy would require a finite number of steps.

> The reason for this procedure to produce approximations for the square root is usually given by calculus, but the procedure itself is known at least from the times of Heron, a Greek mathematician from the first century of CE.

In section 9.11 we will see an example of a procedure to compute π to any desired accuracy. How many such recipes can we formulate using a finite alphabet? The number of recipes is countable. Indeed, the simplest procedure will have just one character, which can be a digit, equal to that number; there are a finite number of those. Then, we list all the procedures that use two characters; there are N_a^2 of those, where N_a is the number of characters in our alphabet. We count those 2-character procedures, then proceed to counting 3-character ones and so on.

But the number of real numbers is *not* countable! This means that some real numbers (in fact, a majority of them) cannot be defined in any way, no matter how hard we try to do that. We simply do not have enough recipes or symbols in a finite alphabet to do that. In some sense, almost all real numbers are irrational, and among those, almost all are not definable *in any way.*

2.8.7 The Completeness of Real Numbers

We already know that rational numbers have "gaps," and we know that irrational numbers fill at least *some* of these gaps. If we take successive truncations of the decimal representation for $\sqrt{2}$—that is, $1, 1.4, 1.41, 1.414, 1.4142, 1.41421, 1.414213, \ldots$, we say that they get ever closer to the irrational number $\sqrt{2}$. However, there is no such *rational* number to which this particular string of decimals gets ever closer.

Above, we defined the real numbers as the set that includes all rational and irrational numbers. This leads to a question: do real numbers have any "gaps," like rational numbers do? Let's take a finite-length, noninteger decimal number and build an infinite sequence by successively adding digits to that number, as in $1.2, 1.23, 1.231, 1.2316, 1.23160, 1.231608, \ldots$. Each next digit will increase the previous value by an ever-smaller amount. Is there a case that there is no such real number, that this sequence of decimals gets ever closer to?

The answer to this question is that real numbers have no "gaps" or "missing points." Every string of decimals defines a real number, and any real number is defined by a string of decimals. This is the foundation of many important statements in calculus.

Another form of this statement is as follows. Take a set $\{x\}$ of real numbers x, such that it has an *upper bound*. This term means that there is such a number $\bar{x}$ that all values of x are less than or equal to $\bar{x}$. For example, define a set $\{x\}$ as all such numbers that $x \leq 2$, then 2 is an upper bound for set $\{x\}$. It is clear that there are multiple upper bounds. Indeed, if $x \leq 2$, then $x \leq 3$, so 3 is also an upper bound of this set. Therefore, set $\{x\}$ defines an infinite set of upper bounds. The completeness postulate states that this set of upper bounds always has the smallest element called the *least upper bound*.

Now let's define a set of real numbers such that $x \leq \sqrt{2}$. This will be an infinite set; some numbers included in this set will be $-100, -1, 0.5, 1, 1.4, 1.41, 1.414, 1.4142, 1.41421$, or 1.414213. It will have upper bounds, for example, 1.5 or 1.45. For real numbers, there is the least upper bound, $\sqrt{2}$.

Compare this to the case of rational numbers. Let's define a set of rational numbers such that $x \leq \sqrt{2}$. In the set of rational numbers, there will be no least upper bound because rational numbers happen to have a "gap" where that least upper bound should be. No matter how close a rational upper bound is to $\sqrt{2}$, there will be another rational upper bound that is even closer. Indeed, let's consider some rational number $r > \sqrt{2}$ and let's assume that decimal representations of r and of $\sqrt{2}$ match for the first M digits, with the $(M+1)$-th digit being different. We always can alter one or more of the digits farther out in the decimal representation of r to make it smaller, yet keep it greater than $\sqrt{2}$. This example shows that while the set of real numbers $x \leq 2$ has a least upper bound, an analogous set of rational numbers does not have one. We will use the existence of a least upper bound in the proof of an important theorem in chapter 9.

Exercises

1. Prove that for any numbers m, n, k, l, it is the case that $(m+n)+(k+l) = (m+l)+(n+k)$. (Note that we have never defined an expression like $m+n+k$ with three or more terms and without parentheses.)
2. (a) Prove that for any numbers m, n, k, l, it is the case that $(m \cdot n) \cdot (k \cdot l) = (m \cdot (n \cdot (k \cdot l)))$. (Note that we have never defined an expression like $m \cdot n \cdot k$ with three or more terms and without parentheses.)
   (b) Extend this reasoning to prove that $(a_1 \cdot (a_2 \cdot a_3)) \cdot (b_1 \cdot (b_2 \cdot b_3)) = (a_1 \cdot (a_2 \cdot (a_3 \cdot (b_1 \cdot (b_2 \cdot b_3)))))$.
3. Prove that $(a+b) \cdot (c+d) = (a \cdot c + b \cdot c) + (a \cdot d + b \cdot d)$.
4. Where in the proof of the fundamental theorem of arithmetic did we use the following concepts and tools: "by definition," mathematical induction, considering all cases, and a constructive proof?
5. Prove that the square of any odd number is odd.
6. Prove that a number is even or odd if its last digit is even or odd.
7. What is the decimal representation for $\frac{10}{7} + \frac{7}{10} = \frac{149}{70}$? (Hint: Use the results from section 2.4.2.)
8. We learned that $1 = 0.9999...$ Give another example of a rational number whose decimal representation is not unique.
9. What condition must satisfy natural numbers m and n in order for $(\sqrt{m} + \sqrt{n})^2$ to be a rational number?
10. Prove that if r is an irrational number, then $1/r$ is also an irrational number.
11. Number e is equal to $e = 2.718281828...$ Based on the first 10 digits, can you conclude if this number is rational or irrational?
12. A computer scientist wrote a code to compute the first million digits of the number called the golden ratio. An analysis of the first million digits detected no periodicity there. Can one conclude that the golden ratio is an irrational number?
13. Prove that $n^3 - n$ is divisible by 6. Provide an algebraic proof and one by induction.
14. Prove by induction that $n^3 + 23n$ is divisible by 6.
15. Prove by induction that $n^5 - 6n$ is divisible by 5.
16. Note that the formulation of theorem 2.19 states that the representation is unique, but theorem 2.20 does not say anything about uniqueness. Why?

17. (a) Prove that the set of nonnegative integers have the same cardinality as the set of all integers.
    (b) Prove that rational numbers have the same cardinality as the set of all integers (that is, not necessarily positive).
18. Prove theorem 2.24.
19. Prove that if the sum of all digits of a natural number is divisible by 3, then this number is divisible by 3. (Hint: start from a two digit number N and represent it as $N = 10n_1 + n_2$, where n_1, n_2 are the digits forming the decimal representation of number N.)
20. Prove by induction that any integer number $n \geq 8$ can be represented as a sum $3k + 5l$, where k, l are integers.
21. Prove that for any integer values of m, n numbers $m^2 + m$ and $mn(m + n)$ are even.
22. Prove that any number $n = 4p$, where p is an integer, is a difference of two squares. (Hint: use $m^2 - k^2 = (m-k)(m+k)$ and set $m - k = 2$.)
23. Prove that any odd number can be represented as a difference of two consecutive squares.
24. Prove that for any integer n the expression $n^2 + 1$ cannot be represented as $4k + 3$, where k is an integer. (Hint: consider two cases: n is odd and n even. In the second case, consider the remainder from dividing $n^2 + 1$ by 4.)
25. Prove that for any real numbers u, v, w, z it is the case that

$$(u - v)(w - z) + (u - z)(v - w) = (u - w)(v - z). \qquad (2.133)$$

    Does it hold for complex numbers as well? Why?
26. Which property of rational numbers was used in the proof of theorems 2.13 and 2.14?
27. In the first proof of theorem 2.10, we used the fact that every natural number can be represented as a product of primes. Do we need to use the fact that such a representation is unique to prove theorem 2.10? If yes, where?
28. Use properties of addition and multiplication to prove that $(-1) \cdot a = -a$.
29. Prove that the place-value representation of two-digit integers given by theorem 2.9 is unique.
30. Write a proof of theorem 2.14. The theorem states that the product of a nonzero rational number and an irrational number is an irrational number. The condition of the rational number being nonzero is important: a product of zero and any irrational number is zero,

which is a rational number. Where is this condition used in the proof?

31. Theorem 2.31 proves that a closed and a half-closed interval have the same cardinality. Prove that an open and a closed interval also have the same cardinality.

32. We stated that the square root of an integer is either an integer or an irrational number (p. 38). Prove this statement by modifying the first proof of theorem 2.10 for the irrationality of $\sqrt{2}$.

3

Inequalities

3.1 Basic Properties

Inequalities are a ubiquitous and useful part of mathematics, from theory to everyday life. We routinely compare values: a product in one store is sold for less than in another store, your neighbor is older than you, and so on. In this chapter, we show that many familiar properties of inequalities can be proved from a few basic ones. We start from several properties that we accept without a proof:

1. Reflexivity: $a \geq a$.
2. Transitivity: if $a \geq b$ and $b \geq c$, then $a \geq c$.
3. Anti-symmetry: if $a \geq b$ and $b \geq a$, then $a = b$.
4. If $a \geq b$, then $a + c \geq b + c$.
5. If $a \geq 0$ and $b \geq 0$, then $ab \geq 0$.
6. For any two numbers a and b it is the case that $a \geq b$ or $b \geq a$.
7. Inequalities $a \geq b$ and $b \leq a$ are equivalent.

We also define *strict inequalities* $a > b$ as such that $a \geq b$ and $a \neq b$. From these basic properties, we can derive many others. These derivations are the subject of the following sections.

Inequalities have many practical applications. They are used for optimization—that is, any time when we seek a smaller or a larger value: a company wants to maximize its profit; a scientist wants to achieve the most accurate measurement, which means minimizing the measurement error; a homeowner refinances her mortgage to reduce payments. In what is sometimes less appreciated, nature often seeks optimal solutions too. Here are two very important examples.

1. Natural selection maximizes the fitness of the species. This process has made us, Homo Sapiens, the species we are.

DOI: 10.1201/9781003456766-3

2. As we shall see below (p. 200), ray reflections from an ellipse minimize the distance traveled by the ray from one focus to another. This is connected to the more general *Fermat's principle*, stating that ray propagation "selects" the path with the shortest travel time, as compared to adjacent paths. A similar principle is true for a particular formulation of mechanical motion, known as the principle of least action.

These phenomena are very different, but both were on the mind of the 18th century French mathematician Pierre Louis Moreau de Maupertuis. He contributed to formulating the principle of least action *and* wrote about the evolution of species in terms of natural selection long before Darwin. (Of course, Darwin's contribution was crucial for the acceptance of this idea because he collected extensive data to support it.)

3.2 Several Theorems

From the basic properties of inequalities, we can derive (prove) many other ones. Here are some examples:

Theorem 3.1 *If $a \geq b$, then $-b \geq -a$.*

Proof
We take $a \geq b$ and add $-a - b$ to both sides.

Theorem 3.2 *If $a \geq b$ and $c \geq d$, then $a + c \geq b + d$.*

Proof
This proof has three steps.

1. *We use property 4. Since $a \geq b$, then $a + c \geq b + c$.*
2. *We again use property 4. Since $c \geq d$, then $c + b \geq d + b$.*
3. *From the last two statements $a + c \geq b + c \geq b + d$ and from the transitivity property 2 we get: $a + c \geq b + d$.*

Theorem 3.3 *If $a \geq b$ and $c \geq 0$, then $ca \geq cb$.*

Proof
This proof has four steps:

1. *We use property 4 and add $-b$ to both sides. Since $a \geq b$, then $a - b \geq 0$.*

2. *We use property 5. Since* $c \geq 0$, *then* $c(a-b) \geq 0$.
3. *Expand:* $ca - cb \geq 0$.
4. *We again use property 4 and add* cb *to both sides. Since* $ca - cb \geq 0$, *then* $ca \geq cb$.

Theorem 3.4 *If* $a \geq b$ *and* $c \leq 0$, *then* $bc \geq ac$.

Proof

At first, we use the previous theorem to prove that $a|c| \geq b|c|$. *Then, we note that for* $c \leq 0$, *we have* $|c| = -c$. *This produces* $-ac \geq -bc$. *Finally, we use theorem 3.1 to flip the sign of the inequality:* $bc \geq ac$.

Theorem 3.5 *If* $a \geq b \geq 0$ *and* $c \geq d \geq 0$, *then* $ac \geq bd$.

Proof

1. *From* $a \geq b$ *and* $c \geq 0$ *we conclude that* $ac \geq bc$.
2. *From* $c \geq d$ *and* $b \geq 0$ *we conclude that* $bc \geq bd$.
3. *From the last two statements and from the transitivity property 2 we obtain* $ac \geq bd$.

Theorem 3.6 *For any* a *it is the case that* $a^2 \geq 0$.

Proof

We have either $a \geq 0$ *or* $a \leq 0$ *(or both, if* $a = 0$*). In the first case, we multiply this inequality by* a, *and the sign of the inequality holds. In the second, it flips. In both cases, we get* $a^2 \geq 0$.

Theorem 3.7 *For* $a > 0$ *and* $b > 1$ *it is the case that*

$$ab > a. \tag{3.1}$$

Proof

If $b > 1$, *then* $b - 1 > 0$. *Let's denote* $c = b - 1$, *where we observe that* $c > 0$. *Then inequality (3.1) is in the form* $a + ac > a$, *which is equivalent to* $ac > 0$. *This last inequality is true due to property 5.*

Theorem 3.8 *For* $a > 0$ *and* $b < 1$ *it is the case that*

$$ab < a. \tag{3.2}$$

Proof

The proof is very similar, but we use $b - 1 < 0$ *this time. Then, we walk through the same line of arguments as in the previous theorem.*

Theorem 3.9 *For* $a > 0$ *it is the case that*

$$\frac{1}{a} > 0. \tag{3.3}$$

Proof
The proof is by contradiction. Suppose that $1/a < 0$. We multiply this inequality by a to get $1 < 0$.

Theorem 3.10 *For $a > b > 0$ it is the case that*

$$\frac{1}{a} < \frac{1}{b}. \tag{3.4}$$

Proof
Due to theorem 3.9 we have $1/(ab) > 0$. We multiply inequality $a > b$ by $1/(ab)$ to get the desired result.

> Mathematics started to study inequalities relatively late. In antiquity, inequalities rarely were the subject of inquiry. One exception was an attempt to approximate π, which is found in many the ancient traditions. The best known is Archimedes' result that $223/71 < \pi < 22/7$. In the decimal system, that would approximately correspond to $3.14084507 < \pi < 3.1428571$.

3.3 Several Inequalities

In this section we prove several inequalities, some of which may seem obvious and some not.

3.3.1 A Sum of Absolute Values

Theorem 3.11 *For any two numbers a and b it is the case that*

$$|a + b| \leq |a| + |b|. \tag{3.5}$$

Proof
Inequality

$$|c| \leq d, \tag{3.6}$$

is equivalent to a double inequality

$$-d \leq c \leq d. \tag{3.7}$$

When saying that inequalities (3.6) and (3.7) are equivalent, we mean that either one can be obtained from the other; this is used below in the proof. From this equivalence and from $|a| \leq |a|, |b| \leq |b|$ we get

$$\begin{aligned} -|a| \leq a \leq |a|, \\ -|b| \leq b \leq |b|. \end{aligned} \tag{3.8}$$

We add these two inequalities to obtain

$$-(|a| + |b|) \leq a + b \leq |a| + |b|. \tag{3.9}$$

We again use the equivalence of inequalities (3.6) and (3.7) to get the desired result:

$$|a + b| \leq |a| + |b|. \tag{3.10}$$

(Note that we used the equivalence of inequalities (3.6) and (3.7) both ways!)

This theorem can also be proved by following the logic in the proofs of theorem 2.27 and setting the imaginary parts of the complex numbers to zeros. It also can be extended to any finite number of terms:

Theorem 3.12 *For numbers $a_1, a_2, \ldots, a_n$ it is the case that*

$$|a_1 + a_2 + \ldots + a_n| \leq |a_1| + |a_2| + \ldots + |a_n|. \tag{3.11}$$

Proof

The proof is by induction.

1. *For two terms, inequality (3.11) holds. This is the statement of theorem 3.11.*
2. *We assume that it holds for n terms and consider the case of $n+1$ terms:*

$$\begin{aligned} |a_1 + a_2 + \ldots + a_n + a_{n+1}| &= |(a_1 + a_2 + \ldots + a_n) + a_{n+1}| \\ &\leq |a_1 + a_2 + \ldots + a_n| + |a_{n+1}| \\ &\leq |a_1| + |a_2| + \ldots + |a_n| + |a_{n+1}|, \end{aligned} \tag{3.12}$$

   where we used theorem 3.11.

3.3.2 The Arithmetic Mean

The arithmetic mean of two numbers a and b is defined as $A = (a + b)/2$.

Theorem 3.13 *For any a and b their arithmetic mean $A = (a + b)/2$ is in between these two numbers. This means that at least one of the following conditions holds:*

$$a \geq \frac{a+b}{2} \geq b \tag{3.13}$$

or:

$$b \geq \frac{a+b}{2} \geq a. \tag{3.14}$$

(For this and other definitions of a mean, the statement "A is between a and b" includes the case when $a = A = b$.)

Proof

It is sufficient to prove that $(A-a)\cdot(A-b) \leq 0$. (Indeed, if $(A-a)$ and $(A-b)$ have different signs, then A is in between a and b. This way of proving theorem 3.13 does not require us to make assumptions on which of the values of a and b is larger.) We use the definition of A to get

$$\begin{aligned}(A-a)\cdot(A-b) &= \left(\frac{a+b}{2}-a\right)\cdot\left(\frac{a+b}{2}-b\right) \\ &= \frac{(a+b-2a)}{2}\cdot\frac{(a+b-2b)}{2} \\ &= \frac{(b-a)}{2}\cdot\frac{(a-b)}{2} \\ &= -\frac{(a-b)^2}{4} \\ &\leq 0.\end{aligned} \tag{3.15}$$

3.3.3 The Geometric Mean

For any two nonnegative numbers a and b, the *geometric mean* is defined as $G = \sqrt{ab}$.

Theorem 3.14 *For any nonnegative a and b, it is the case that either $a \leq \sqrt{ab} \leq b$, or $b \leq \sqrt{ab} \leq a$.*

Proof

Similarly to the previous theorem, we prove that $(G-a)(G-b) \leq 0$. We use the definition of G:

$$G = (\sqrt{ab}-a)\cdot(\sqrt{ab}-b) \leq 0. \tag{3.16}$$

We observe that

$$\begin{aligned}\sqrt{ab}-a &= \sqrt{a}(\sqrt{b}-\sqrt{a}), \\ \sqrt{ab}-b &= \sqrt{b}(\sqrt{a}-\sqrt{b}).\end{aligned} \tag{3.17}$$

This yields

$$\begin{aligned}G &= \sqrt{ab}(\sqrt{b}-\sqrt{a})(\sqrt{a}-\sqrt{b}), \\ &= -\sqrt{ab}(\sqrt{b}-\sqrt{a})^2 \\ &\leq 0.\end{aligned} \tag{3.18}$$

3.3.4 The Harmonic Mean

For any two positive numbers a and b, the *harmonic mean* is defined as

$$H = \frac{2}{\frac{1}{a} + \frac{1}{b}}. \tag{3.19}$$

Theorem 3.15 *For a positive a and b one of the following inequalities holds: $a \geq H \geq b$ or $b \geq H \geq a$.*

Proof

The harmonic mean is equal to

$$H = \frac{2ab}{a+b}. \tag{3.20}$$

Similarly to the previous theorems, we prove that $(H - a)(H - b) \leq 0$. We substitute here the explicit expressions for H, compute common denominators for the terms in parentheses, and factor the numerators:

$$\begin{aligned}
(H-a)\cdot(H-b) &= \left(\frac{2ab}{a+b} - a\right)\cdot\left(\frac{2ab}{a+b} - b\right) \\
&= \left(\frac{ab-a^2}{a+b}\right)\cdot\left(\frac{ab-b^2}{a+b}\right) \\
&= \left(\frac{a(b-a)}{a+b}\right)\cdot\left(\frac{b(a-b)}{a+b}\right) \\
&= -\frac{ab(b-a)^2}{(a+b)^2} \\
&\leq 0.
\end{aligned} \tag{3.21}$$

3.3.5 The Quadratic Mean

For any two nonnegative numbers a and b, the *quadratic mean* is defined as

$$Q = \sqrt{\frac{a^2+b^2}{2}}. \tag{3.22}$$

Theorem 3.16 *For nonnegative a and b it is the case that $a \geq Q \geq b$ or $b \geq Q \geq a$.*

Proof

From theorem 3.5 it follows that for any nonnegative numbers p and q, the inequality $p \geq q$ holds if and only if $p^2 \geq q^2$. The statement of the theorem is equivalent to $a^2 \geq Q^2 \geq b^2$ or $b^2 \geq Q^2 \geq a^2$. We observe that Q^2 is the arithmetic mean of a^2 and b^2. Then the desired result follows from theorem 3.13.

3.3.6 Inequalities for the Four Means

Theorem 3.17 *For any positive a and b, the following inequalities hold:*

$$H \le G \le A \le Q. \tag{3.23}$$

Proof

1. *First, we compare the geometric and arithmetic means*

$$\frac{a+b}{2} \ge \sqrt{ab}. \tag{3.24}$$

We add $-\sqrt{ab}$ to both sides and compute a common denominator:

$$\frac{a+b-2\sqrt{ab}}{2} \ge 0. \tag{3.25}$$

Here $a = (\sqrt{a})^2$ and $b = \left(\sqrt{b}\right)^2$. Then the value in the numerator is a complete square:

$$\frac{(\sqrt{a})^2 - 2\sqrt{ab} + \left(\sqrt{b}\right)^2}{2} = \frac{\left(\sqrt{a}-\sqrt{b}\right)^2}{2} \ge 0. \tag{3.26}$$

The left-hand side is always nonnegative, which proves $G \le A$. (Inequality $G \le A$ is used quite often. We will refer to it in sections 5.9 and 7.18.)

2. *Next, we compare the harmonic and the geometric means. We must prove that*

$$\frac{2}{\frac{1}{a}+\frac{1}{b}} \le \sqrt{ab}. \tag{3.27}$$

Computing the common denominator produces

$$\frac{2ab}{a+b} \le \sqrt{ab}. \tag{3.28}$$

Next, we multiply both sides by $a+b$ and divide by $2\sqrt{ab}$ to get:

$$\sqrt{ab} \le \frac{a+b}{2}. \tag{3.29}$$

This is the previous inequality $G \le A$, which we have proved already.

3. *Finally, we prove* $A \leq Q$*:*

$$\frac{a+b}{2} \leq \sqrt{\frac{a^2+b^2}{2}}. \tag{3.30}$$

Squaring both sides yields

$$\frac{a^2+b^2+2ab}{4} \leq \frac{a^2+b^2}{2}. \tag{3.31}$$

We collect the terms to get

$$0 \leq \frac{a^2+b^2-2ab}{4}. \tag{3.32}$$

The right-hand side contains a complete square $a^2+b^2-2ab = (a-b)^2$*, which is nonnegative.*

3.3.7 Bernoulli's Inequality

The following inequality bears the name of Jacob Bernoulli.

Jacob Bernoulli, a member of a family of mathematicians is known for several important contributions, including those to probability theory.

When we flip a coin, we generally cannot predict which side it will land on. This is an example of a process that is modeled in mathematics as a so-called *random event.* However, if we flip a coin many times, for example, 100 or 1,000, one thing becomes predictable: the outcomes are typically split approximately evenly between heads and tails. Similar trends are observed for other random events. Probability theory, a branch of mathematics that studies random events, calls this phenomenon *the law of large numbers.* The credit for formulating this law goes to Jacob Bernoulli.

Probabilities of getting a particular number of heads (or tails) in a series of coin flips are expressed through the binomial coefficients, which we will explore in section 5.2. If the number of coin flips in the series is large, these probabilities can also be approximated by the so-called bell curve. This model uses the exponential function, which we will discuss in section 6.5. This is an example of how different mathematical concepts are linked together.

Theorem 3.18 *For any* $p \geq -1$ *and any integer* $n > 0$ *it is the case that*

$$(1+p)^n \geq 1+pn. \tag{3.33}$$

Proof

The proof is by induction:

1. *For* $n = 1$*:*

$$(1+p)^1 \geq 1+p, \tag{3.34}$$

which is true.

2. *We use the inequality for* n *and multiply it by* $(1+p)$*:*

$$(1+p)^{n+1} \geq 1+np+p+np^2. \tag{3.35}$$

Since $p^2 \geq 0$ *and* $n > 0$*, the right-hand side here satisfies the following inequality:*

$$1+np+p+np^2 > 1+(n+1)p. \tag{3.36}$$

From the last two inequalities we get the statement of the theorem for $(n+1)$*.*

3.3.8 The Cauchy-Schwarz Inequality

To formulate this theorem we introduce the so-called *sigma notation* for sums. It is used in cases when the number of terms in a sum may vary or when there are too many terms to be listed explicitly. For example, we need to use the sum of N indexed values $a_1 + a_2 + \ldots + a_N$. The sigma notation allows us to concisely write such a sum as

$$a_1 + a_2 + \ldots + a_N = \sum_{n=1}^{N} a_n, \tag{3.37}$$

where the subscript and superscript for the Greek letter sigma (Σ) show the range of indices n for the terms a_n in the sum. Using this notation, the Cauchy-Schwarz inequality is given by the following theorem:

Theorem 3.19 *For two sets of numbers,* $x_1, x_2, \ldots, x_N$ *and* $y_1, y_2, \ldots, y_N$ *it is the case that*

$$\left(\sum_{n=1}^{N} x_n y_n\right)^2 \leq \left(\sum_{n=1}^{N} x_n^2\right) \cdot \left(\sum_{n=1}^{N} y_n^2\right). \tag{3.38}$$

Proof

This inequality is not easy to prove. Let's start from a simpler case $N = 2$*. Then the inequality is written as*

$$(x_1 y_1 + x_2 y_2)^2 \leq \left(x_1^2 + x_2^2\right) \cdot \left(y_1^2 + y_2^2\right). \tag{3.39}$$

We expand the square in the left-hand side and the product in the right-hand side to get

$$x_1^2 y_1^2 + x_2^2 y_2^2 + 2x_1 y_1 x_2 y_2 \leq x_1^2 y_1^2 + x_2^2 y_2^2 + x_2^2 y_1^2 + x_1^2 y_2^2. \tag{3.40}$$

The first two terms in the left-hand side and right-hand side cancel. We move the remaining term in the left-hand side to the right to get:

$$0 \leq x_2^2 y_1^2 + x_1^2 y_2^2 - 2x_1 y_1 x_2 y_2. \tag{3.41}$$

In the right-hand side, we recognize a complete square:

$$0 \leq (x_2 y_1 - x_1 y_2)^2 . \tag{3.42}$$

The square in the right-hand side is nonnegative, which proves the inequality.

For $N > 2$ a proof is more difficult. It illustrates a situation when a proof starts from a seemingly unrelated point and then somehow the desired result pops up. We define a function $F(t)$ as follows:[1]

$$F(t) = \sum_{n=1}^{N} (x_n + t y_n)^2 . \tag{3.43}$$

This function is a sum of squares and therefore is nonnegative for all values of t:

$$F(t) \geq 0. \tag{3.44}$$

Next, we expand the squares in the definition of $F(t)$ and regroup the terms:

$$\begin{aligned} F(t) &= \sum_{n=1}^{N} (x_n + t y_n)^2 \\ &= \sum_{n=1}^{N} \left(x_n^2 + t^2 y_n^2 + 2t x_n y_n\right) \\ &= \sum_{n=1}^{N} x_n^2 + t^2 \sum_{n=1}^{N} y_n^2 + 2t \sum_{n=1}^{N} x_n y_n. \end{aligned} \tag{3.45}$$

We denote

$$\begin{aligned} C &= \sum_{n=1}^{N} x_n^2, \\ A &= \sum_{n=1}^{N} y_n^2, \\ B &= 2 \sum_{n=1}^{N} x_n y_n. \end{aligned} \tag{3.46}$$

Then

$$F(t) = At^2 + Bt + C, \tag{3.47}$$

[1] We will talk more about functions in chapter 4.

which indicates that $F(t)$ is a quadratic function of t. We already know that this function in nonnegative. This means that the equation

$$At^2 + Bt + C = 0, \tag{3.48}$$

must either have no real roots, or both of its roots must be equal to zero. Then the discriminant of this quadratic equation must be nonpositive (see theorem 5.5):

$$B^2 - 4AC \leq 0. \tag{3.49}$$

We recall the definitions of A, B, and C to get

$$4\left(\sum_{n=1}^{N} x_n y_n\right)^2 - 4\left(\sum_{n=1}^{N} x_n^2\right) \cdot \left(\sum_{n=1}^{N} y_n^2\right) \leq 0. \tag{3.50}$$

After dividing this inequality by 4 and moving the second term to the right-hand side we get the desired Cauchy-Schwarz inequality.

What is the condition for having the equal sign in the Cauchy-Schwarz inequality? It is not obvious from the inequality itself, but can be deduced from the starting point of the proof—that is, from the definition of function $F(t)$ in equation (3.43). Indeed, the proof is based on $F(t) \geq 0$, and the equal sign for the Cauchy-Schwarz inequality can be traced to $F(t) = 0$. Since $F(t)$ is a sum of nonnegative terms, in order for $F(t)$ to be 0, there must exist such a value of t that $x_n = -ty_n$ for all n. This condition has a geometric interpretation: if x_n and y_n are the components of a vector, these two vectors must be colinear. This is an example of how details of a proof provide an additional insight into properties of a mathematical object.

Exercises

1. Prove that if $a \geq b$ is false, then $a < b$.
2. For two positive rational numbers $r_1 = m/n$ and $r_2 = k/l$ prove that $(m + k)/(n + l)$ is in between r_1 and r_2.
3. Prove that for any $a \neq 0$ it is the case that $|a + \frac{1}{a}| \geq 2$.
4. Prove that for any $a \neq 0$ it is the case that $|a + \frac{1}{2a}| \geq \sqrt{2}$.
5. Detect a pattern in the inequalities in exercises 3 and 4. Use that pattern to generalize them and prove the generalized result.
6. Prove that for any a, b it is the case that $a^2 + ab + b^2 \geq 0$.
7. Prove the triangle inequality: $|x - y| \leq |x - z| + |z - y|$.

8. A student is solving the following inequality:

$$\sqrt{x} < (c+d)(c-d) - c^2. \tag{3.51}$$

She squares this inequality to get rid of the square root:

$$\begin{aligned} x &< \left((c+d)(c-d) - c^2\right)^2 \\ &= (c+d)^2(c-d)^2 - 2c^2(c+d)(c-d) + c^4 \\ &= c^4 - 2c^2d^2 + d^4 - 2c^4 + 2c^2d^2 + c^4 \\ &= d^4. \end{aligned} \tag{3.52}$$

Why is this solution incorrect?

9. Prove that $|u - v| \geq |u| - |v|$.

10. Prove that

$$(a+b)(1/a+1/b) \geq 4$$

for a positive a, b. Provide two proofs: one using the Cauchy-Schwarz inequality and another without it.

11. Prove that

$$(a+b+c)(1/a+1/b+1/c) \geq 9$$

for nonzero a, b, c.

12. Generalize the inequalities in exercises 10 and 11 and prove the generalized result.

13. Show that the inequality in exercise 3 is a particular case of the inequality in exercise 10.

14. A proof of irrationality of $\sqrt{2}$ starting on p. 37 uses the fact that $1 < \sqrt{2} < 2$. Prove this inequality.

15. Prove that if A is the arithmetic mean of two nonnegative numbers a and b and $A = a$, then $A = b$. Is the same true for the geometric and quadratic means?

16. Theorem 3.18 is formulated for $p \geq -1$. Is it valid for $p < -1$? If not, at which step is the condition $p \geq -1$ used in the proof of this theorem?

17. Provide alternative proofs for theorems 3.13, 3.14, and 3.15. Start from the statement "without loss of generality we assume $a \leq b$."

18. Where is theorem 3.5 used in the proof of theorem 3.17?

19. In section 2.8.7 we learned the concept of the upper bound and that any real number defines an infinite set of upper bounds. Consider two numbers, a and b, such that $a < b$, and the corresponding sets of upper bounds, S_a and S_b. Which of these two sets is a subset of the other?
20. Prove that if $a, b > 1$, then $ab - (a + b) + 1 > 0$.
21. Prove that for any $a > b$ it is the case that $a^3 - b^3 \geq a^2b - ab^2$. What is the condition for having the equal sign in this inequality?
22. Prove that for any $a, b \geq 0$ it is the case that $a^3 + b^3 \geq a^2b + ab^2$. What is the condition for having an equal sign in this inequality? Is it different from that in exercise 21?
23. Prove that for any $a \neq 0$ it is the case that

$$a^2 + \frac{1}{a^2} \geq a + \frac{1}{a}. \tag{3.53}$$

24. Prove by induction that for any positive a, b, such that $a \neq b$, and for any nonnegative integer k it is the case that

$$\frac{a^n - b^n}{a - b} \geq n\sqrt{a^{n-1}b^{n-1}}, \tag{3.54}$$

where $n = 2^k$. For what values of a, b, and k are the two sides of this inequality equal? (Hint: note that the numerator in the left-hand side is a difference of squares.)

25. Generalize the result of exercise 24 to any natural n—that is, assume that n is not necessarily a power of 2. (Hint: use induction, the identity $(a + b)(a^n - b^n) + (a - b)(a^n + b^n) = 2(a^{n+1} - b^{n+1})$ and the inequality for the arithmetic and geometric means.)
26. Prove that for any $a, b \neq 0$ it is the case that

$$\frac{ab}{a^2 + ab + b^2} \leq \frac{1}{3}. \tag{3.55}$$

Use two methods:

(a) use the inequality for the arithmetic and geometric means (theorem 3.17) and

(b) use theorem 2.19.

What is the condition on a and b to have the equal sign in this inequality?

27. Prove that for any a it is the case that

$$\frac{a}{a^2 + 3a + 4} \leq \frac{1}{7}. \tag{3.56}$$

What is the condition on a to have the equal sign in this inequality? (Hint: check the sign of the discriminant for the expression in the denominator, or use $a^2 + 3a + 4 > 0$ without proof.)

28. Are the following statements true for all values of the variables? (Assume that denominators, if present, are nonzero.)

    (a) If $a > 3$, then $9/a < 3$.
    (b) If $a > b$, then $b^2/a < b$.
    (c) If $|a| > b$, then $a > b$ or $a < -b$.
    (d) If $3/(a-1) > 1$, then $3 > a - 1$.
    (e) If $3/(a-1) < 1$, then $3 < a - 1$.

29. Consider the following sum:

$$S(n) = q^{-n} + q^{-n+2} + \ldots + q^{n-2} + q^n, \tag{3.57}$$

where n is an even natural number and $q > 0$. Use induction and the result from exercise 3 to prove that $S(n) \geq n + 1$.

30. What is the condition for having $|a + b| = |a| + |b|$? Can you trace the case of $|a + b| = |a| + |b|$ to inequalities in the proof of theorem 3.11?

31. Prove inequality (3.24) for the geometric and arithmetic mean from the Cauchy-Schwarz inequality (3.38).

4

Functions

4.1 Definition and Examples

A function is defined by three things:[1]

1. A set of numbers called the *domain*.
2. A set of numbers called the *range*.
3. For every number in the domain, a way to associate one and only one number in the range.

Note that if we associate two or more numbers in the range to a number in the domain, that will not be a function. Also note that there are no restrictions on how to define the two sets as long as they allow the association between them.

Sometimes we will omit definitions of the domain and the range. In these cases, we will assume that both are the largest subsets of real numbers that are possible within the context of that function definition. For example, we may say that a function is defined by $f(x) = 1/(x - 1)$ without specifying the domain and the range. This means that the domain is all real numbers, except $x = 1$, and the range is all real numbers, except $f = 0$.

Finally, note that there are no restrictions on how to define the association, except as stated above. Common examples of functions may create the impression that it should be some kind of algebraic formula. This is not a requirement; the recipe for association can be anything.

> Today, the ubiquitous concept of the plot of a function is almost synonymous with the definition of one. We imagine the horizontal and vertical axes and points or a curve that visually shows the correspondence between these axes. The first clear description of a plotting procedure was probably done by Oresme (1323–1382), a scholastic philosopher who taught in the University of Paris. He described the arrangement which today we might call a bar graph.

[1] All functions in this book deal with numbers, even though a general definition may include other objects.

DOI: 10.1201/9781003456766-4

Here are several examples of functions that illustrate the above definition:

1. For all real numbers x, function $f(x)$ is computed as $f(x) = 2x + 3$.

   This is a ubiquitous linear function. Its domain is all real numbers, and its range is all real numbers. The association is an algebraic formula that links the two sets.

2. For all real numbers $x \neq 5$, function $f(x)$ is computed as $f(x) = 2x + 3$. The domain of this function is all real numbers, except number 5. The range is all real numbers except number 13. Note that this function is different from the one defined in example 1.

3. For all real numbers x, function $f(x)$ is computed as $f(x) = 0$ if $x < 0$, $f(x) = \frac{1}{2}$ if $x = 0$ and $f(x) = 1$ if $x > 0$. This function is called the step function. The domain of this function is all real numbers. The range contains just three numbers: 0, $\frac{1}{2}$, and 1.

4. For all real numbers x, function $f(x)$ is computed as $f(x) = 1$ if x is a rational number and $f(x) = 0$ if x is an irrational number. The domain of this function is all real numbers. The range contains just two numbers: 0 and 1.

5. The domain of function $f(x)$ contains just one point, $x = 0.33$. The value of the function at this point is $f(x) = 10^{765764}$.

At least since the time of Isaac Newton, a large number of applications of mathematics uses functions. There are multiple reasons for this.

Many natural and socioeconomic phenomena are not stationary, which means that their numerical measure is a function of time: planets orbit the Sun, stock prices vary daily, chemical reactions run out of reagents and stop, and so on.

In addition, many variables vary spatially: temperature is different in different regions of the globe, a radio wave signal attenuates at large distances from the transmitter, and a crop yield is different at different parts of the field. These variables are modeled as functions of coordinates.

The third large set of applications of functions deals with inference: the earning potential of an individual depends on her level of education; life insurance premiums depend on whether the person is a smoker; a particular signal received by a radar means that there is a plane approaching the airport.

Sometimes an algebraic expression for a function contains two or more different numbers that are denoted with letters. For example, we may have a linear function $f(x) = px + q$. For the most part, we are interested in how the function depends on only one number; in the above example that number is denoted by x. This number is called the variable or the argument for that function. The other values (those denoted by p and q) are called *parameters*.

They are assumed to be placeholders for some other numbers, but we do not consider the function's behavior with respect to these parameters. Usually they are held constant.

4.2 Odd and Even Functions

Sometimes a function has a symmetric domain: for any x that belongs to the domain, the value $-x$ also belongs to the domain. Among such functions, we can define two function types that have special properties:

1. An *odd* function is such that for any x in its domain, it is the case that $f(-x) = -f(x)$.
2. An *even* function is such that for any x in its domain, it is the case that $f(-x) = f(x)$.

Of course, many functions are neither even nor odd because the above identities do not hold for them. However, starting from a function with a symmetric domain, we can always define another function that will be odd or even. Suppose we have a function $f(x)$ that has a symmetric domain. We can construct the following two functions:

$$\begin{aligned} f_1(x) &= \frac{f(x)+f(-x)}{2}, \\ f_2(x) &= \frac{f(x)-f(-x)}{2}. \end{aligned} \tag{4.1}$$

Function $f_1(x)$ is even and function $f_2(x)$ is odd. Indeed, let's compute $f_1(-x)$ and $f_2(-x)$:

$$\begin{aligned} f_1(-x) &= \frac{f(-x)+f(x)}{2} = f_1(x), \\ f_2(-x) &= \frac{f(-x)-f(x)}{2} = -f_2(x). \end{aligned} \tag{4.2}$$

This observation is key for proving the following:

Theorem 4.1 *Every function with a symmetric domain can be uniquely represented as a sum of an odd and an even function.*

Proof

First we prove that a function can be represented as a sum of an odd and an even function. The proof is by construction. We start from an arbitrary function $f(x)$ and define functions $f_1(x)$ and $f_2(x)$ by equations (4.1). We *already know that function $f_1(x)$ is even and function $f_2(x)$ is odd. The direct substitution yields*

$$f(x) = f_1(x) + f_2(x). \tag{4.3}$$

This proves the existence of the representation we are seeking.

Next, we prove that this representation is unique. Since we already know that at least one such representation exists, we can write

$$f(x) = f_e(x) + f_o(x), \tag{4.4}$$

where $f_e(x)$ is even and $f_o(x)$ is odd. At this point we do not know if $f_e(x)$ must be identically equal to $f_1(x)$ and $f_o(x)$ must be identically equal to $f_2(x)$. If these identities hold, the representation is unique. We rewrite equation (4.4) for $-x$:

$$f(-x) = f_e(-x) + f_o(-x). \tag{4.5}$$

Next, we add equations (4.4) and (4.5) to get:

$$f(x) + f(-x) = f_e(x) + f_e(-x) + f_o(x) + f_o(-x). \tag{4.6}$$

Here, we observe that $f_o(x) + f_o(-x) = 0$, and $f_e(x) + f_e(-x) = 2f_e(x)$. This yields

$$f_e(x) = \frac{f(x) + f(-x)}{2}. \tag{4.7}$$

Then $f_e = f_1$, with the latter defined by the first equation in (4.1). Similarly, we subtract equations (4.4) and (4.5) to prove that $f_o = f_2$. Thus, any representation of function $f(x)$ as a sum of an even and an odd function is equivalent to that given by equations (4.1) and is therefore unique.

4.3 Composite Functions

Consider two functions, $f(x)$ and $g(y)$. Suppose that the range of $f(x)$ is a subset of the domain of $g(y)$ (in practice, the range of $f(x)$ is often considered to be the same as the domain of $g(y)$). Then, we can define a third function $h(x)$:

1. Its domain is the same as the domain of $f(x)$.
2. To compute function $h(x)$ for any number x in its domain, we do the following:
   (a) We compute $f(x)$. We can do this because the domains of $f(x)$ and of $h(x)$ are the same.
   (b) We use the result $y = f(x)$ as an input to $g(y)$. We can do this because the domain of $g(y)$ includes the range of $f(x)$.

   The result of this computation is the value of the function $h(x)$.

We say that $h(x)$ is a *composite* function. In plain English, we take a number, plug it into function $f(x)$ and then plug the result into function $g(y)$. This is the recipe for computing $h(x)$. We denote this new function as $h(x) = g(f(x))$. Here are two examples:

1. We use:

$$\begin{aligned} f(x) &= 2x + 3, \\ g(y) &= 7y - 2. \end{aligned} \tag{4.8}$$

   The domain and range for each span the entire real axis. Then

$$h(x) = g(f(x)) = g(2x + 3) = 7 \cdot (2x + 3) - 2 = 14x + 19. \tag{4.9}$$

2. We use:

$$\begin{aligned} f(x) &= \sqrt{x}, \\ g(y) &= y^2. \end{aligned} \tag{4.10}$$

   The domain and the range for $f(x)$ is given by $x \geq 0; f(x) \geq 0$. Then

$$h(x) = \left(\sqrt{x}\right)^2 = x. \tag{4.11}$$

   One may look at this formula and decide that since $h(x) = x$, this function is defined for all real values of x. This is incorrect. This function is only defined for $x \geq 0$ because it retains the domain of $f(x)$. Its range is then given by $h(x) \geq 0$.

We already discussed how functions are used to model a dependent relationship between different phenomena or quantities (p. 82). Sometimes the chain of interdependence includes three or more phenomena. For example, the trajectory of a satellite depends on the gravitational force at its current position, and the current position in turn depends on time, as the satellite is constantly moving. This is an example when using a composite function becomes a powerful tool to compute the satellite trajectory: we define $F(r(t))$, where t is time, $r(t)$ is the satellite position as a function of time, and F is the gravitational force at location r. In another example, the ticket sales for an airline depend on the airfare, which in turn depends on the price of fuel.

4.4 Monotonic Functions

Suppose we compute the value of a function at two different values,[2] x_1 and x_2, and suppose that $x_1 > x_2$. What can we say about $f(x_1)$ and $f(x_2)$? Intuitively, we have a concept of an increasing function that will have $f(x_1) > f(x_2)$, and a decreasing function will have $f(x_1) < f(x_2)$. These two notions are formalized in the concept of *monotonic* functions.

If for any pair x_1 and x_2 from the domain, such that $x_1 > x_2$, the inequality $f(x_1) \geq f(x_2)$ is true, such a function is called monotonically increasing. Conversely, if for any x_1 and x_2 from the domain, such that $x_1 > x_2$, the inequality $f(x_1) \leq f(x_2)$ is true, such a function is called monotonically decreasing.

A more narrow statement defines *strictly monotonic functions*, for which we use strict inequalities in the definition. That is, if $f(x)$ is strictly monotonically increasing, then for any pair of numbers $x_1 > x_2$ it is the case that $f(x_1) > f(x_2)$.

Consider monotonic functions $f_i(x), f_d(x), g_i(y)$, and $g_d(y)$. Subscript i means that a function is increasing, and subscript d means that it is decreasing. Now consider the following composite functions:

1. $f_i(g_i(y))$
2. $f_i(g_d(y))$
3. $f_d(g_i(y))$
4. $f_d(g_d(y))$

Theorem 4.2 *Functions $f_i(g_i(y))$ and $f_d(g_d(y))$ are monotonically increasing, and functions $f_i(g_d(y))$ and $f_d(g_i(y))$ are monotonically decreasing.*

Proof

We prove this theorem for function $f_i(g_i(y))$. Consider two values of y such that $y_1 > y_2$. Then $g_i(y_1) > g_i(y_2)$. Used as arguments of function f_i, they yield $f_i(g_i(y_1))) > f_i(g_i(y_2))$.

Proofs for the other three functions are analogous.

4.5 Inverse Functions

If we have some function $f(x)$, sometimes we can define a new function that is called the *inverse* of $f(x)$. Such a function is denoted as $f^{-1}(y)$ and is defined by the following equation:

$$f^{-1}(f(x)) = x, \tag{4.12}$$

[2]Obviously, for this to be possible, the domain of the function must contain more than one point.

for all values of x in the domain of $f(x)$. Here, we see a use of the already familiar composite function. The original function $f(x)$ takes some value of x and computes some value of $y = f(x)$. The inverse function takes that value of y and "restores" the original value of x.

Numerous applications of mathematics require solving equations. Often an engineer or an economist has some data, and from that data and an applicable mathematical model she needs to infer a quantity that was not measured directly.

Look at one of the applications of functions on p. 82. Engineers know that a cell phone signal attenuates as the signal propagates farther from a transmitter. They define a function $S(x)$, where x is the distance from the transmitter, and S is the signal strength. They also have the data indicating that if the signal strength drops below some threshold S_0, a cell phone is no longer able to process that signal.

In a mathematical model, if a phone is close to the cell tower, where the signal is strong, it works, and if the phone is very far, where the signal is severely attenuated, the service drops. In this model there should be a maximum distance at which the phone can still work. How do engineers compute that maximum distance? Such a distance would be the solution of equation $S(x_0) = S_0$, and x_0 can be found here by applying the inverse function to both sides of this equation: $x_0 = S^{-1}(S_0)$. This computation would be important because it would define how far the cell phone company may place cell towers from each other and still maintain a coverage in the area.

The notation $f^{-1}(y)$ for the inverse function has nothing to do with computing a reciprocal value $\frac{1}{f(x)}$. We can partially rationalize this notation if we consider a linear function: $f(x) = ax$. The inverse of this function is given by $f^{-1}(y) = a^{-1}y$, where a^{-1} *is* a reciprocal of a. Indeed:

$$f^{-1}(f(x)) = a^{-1}f(x) = a^{-1}ax = x. \tag{4.13}$$

However, the inverse is different from the reciprocal for all other functions.

From the above definition for composite functions, we conclude that:

1. The domain of the inverse function f^{-1} is the same as the range of function f.

2. The range of the inverse function f^{-1} is the same as the domain of function f.

When does an inverse function exist? We need to go back to the definition of a function, which says that there must be one and only one value of $f(x)$ for each value of x. Both plots in figure 4.1 show that there is one value of $f(x)$ for each value of x. Just trace the dashed lines from x_1 or x_2 up and to the left, to get values of f_1 or f_2.

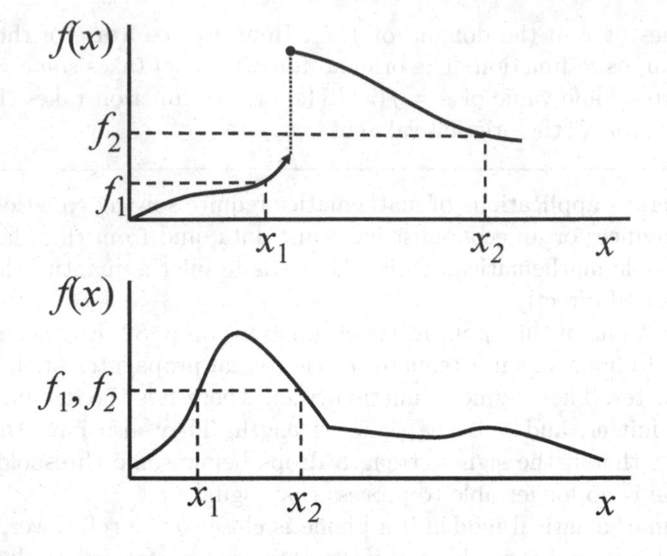

FIGURE 4.1
Conditions for the existence of an inverse function

To recover x_1 or x_2 from f_1 or f_2, we would have to trace the dashed lines from f_1 or f_2 to the right and then down. This works for all values of x, f in the top plot, but does not work for at least some values of x, f in the bottom plot. This means that the top function $f(x)$ has an inverse, and the bottom function does not.

The above criterion is called the *horizontal line test.* While the graphical argument is intuitive, it is a poor one from a mathematical point of view. Graphs are never precise, so we cannot know for sure if the test works. Consider the following function:

$$w(x) = \begin{cases} x, & \text{if } x \text{ is rational.} \\ x+1, & \text{otherwise.} \end{cases} \tag{4.14}$$

Admittedly, this is a weird function that is unlikely to have practical applications, but it illustrates the problem with using graphics for the horizontal line test. A "plot" of this function is shown in figure 4.2. It shows two dotted lines that notionally represent infinite number of holes for rational and, separately, for irrational values of x.

We cannot use this plot to apply the horizontal line test because the holes remain infinitely small, no matter how much we zoom in on the plot. Since each dotted line looks dense, we might assume that this function fails the horizontal line test. Yet, this assumption would be incorrect: $w(x)$ has an inverse. For any irrational x, the value of $w(x)$ is also irrational, and for any rational x, the value of $w(x)$ is rational.[3] The two cases remain well separated.

[3]A proof of this fact is the subject of exercise 29 for this chapter. Also, compare function

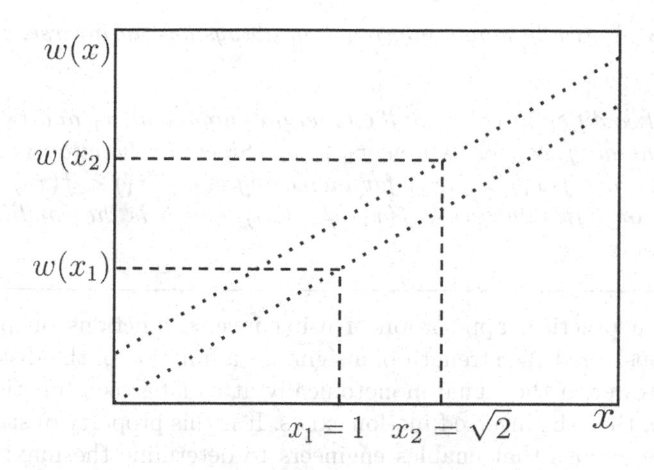

FIGURE 4.2
A weird function

This yields the following formulation for the inverse:

$$w^{-1}(y) = \begin{cases} y, & \text{if } y \text{ is rational,} \\ y - 1, & \text{otherwise.} \end{cases} \tag{4.15}$$

In a less exotic example, consider functions that are defined for all real numbers, so that we can never plot the entire function. Suppose we plot $g(x) = x^3$ for some range of values of x. How do we know if the test works beyond the region that we have plotted? There is no way to make a conclusion from the plot. A determination of the existence of an inverse function must be proved rigorously, and a graphical form of the horizontal line test is not a rigorous argument. The proper criterion for the existence of an inverse is as follows:[4]

Function $f(x)$ has an inverse $f^{-1}(x)$ if and only if for any pair of distinct values $x_1 \neq x_2$ from its domain it is the case that $f(x_1) \neq f(x_2)$.

Here are a few examples:

1. Function $f(x) = x^2$ defined on the domain $x \geq 0$ has an inverse function $f^{-1}(y) = \sqrt{y}$. Indeed, $f^{-1}(f(x)) = \sqrt{x^2} = x$.
2. Function $f(x) = x^2$ defined on the domain of all real values of x does not have an inverse.
3. Function $f(x) = ax + b$ has an inverse $f^{-1}(y) = (y - b)/a$, as long as $a \neq 0$. Indeed, $f^{-1}(f(x)) = ((ax + b) - b)/a = x$.

The existence of the inverse is connected with the notion of monotonic functions.

$w(x)$ with the tangent function $\tan\theta$, which has a curious property: for any rational nonzero θ the value of $\tan\theta$ is irrational (p. 160).

[4]Technically, this criterion misses all functions whose domain is just one number. Such functions do have an inverse, but they are of little practical interest.

Theorem 4.3 *A strictly monotonic function always has an inverse.*

Proof

Consider any two different values of the function's argument, x_1 and x_2. Without loss of generality we can assume $x_1 > x_2$. Since the function is strictly monotonic, we have $f(x_1) > f(x_2)$ for increasing and $f(x_1) < f(x_2)$ for decreasing functions. In either case $f(x_1) \neq f(x_2)$, which is the condition for having an inverse.

> Look at a practical application of using inverse functions on p. 87, where we considered the strength of a signal as a function of the distance from a cell tower. If the signal monotonically attenuates as a function of the distance, then the inverse function exists. It is this property of strictly monotonic functions that enables engineers to determine the maximum workable distance to the cell tower. (In practice, the physics is more complex: the signal strength is not always a monotonic function of the distance because it is affected by the terrain and the buildings.)

This theorem states a sufficient, but not a necessary condition for the existence of an inverse. There are functions that are not monotonic, and yet they have inverses. One example is the function given by equation (4.14). Another example, a more practical one, is $f(x) = 1/x$. While this function is monotonically decreasing for $x > 0$ and, separately, for $x < 0$, we also have $f(x_1) > f(x_2)$ if $x_1 > 0 > x_2$.

The inverse of a strictly monotonic function has a property that we will use later:

Theorem 4.4 *If a function is strictly monotonically increasing, then its inverse is also strictly monotonically increasing.*

Proof

A proof is by contradiction. We use values $f(x_1), f(x_2)$ as inputs to the inverse and assume that the inverse is not strictly monotonically increasing. This means that there exist such values of $f(x_1), f(x_2)$ that (without loss of generality) $f(x_1) > f(x_2)$ and

$$f^{-1}(f(x_1)) \leq f^{-1}(f(x_2)). \tag{4.16}$$

From the definition of the inverse, this inequality can be written as

$$x_1 \leq x_2. \tag{4.17}$$

Combined with the earlier inequality $f(x_1) > f(x_2)$, this contradicts the assumption that $f(x)$ is strictly monotonically increasing.

Similarly, the inverse of a strictly monotonically decreasing function is strictly monotonically decreasing.

4.5.1 Inverse of an Inverse

Suppose that we have a function $f(x)$ that has an inverse $f^{-1}(y)$. What is the inverse of $f^{-1}(y)$?

Theorem 4.5 *If $f^{-1}(y)$ is the inverse of $f(x)$, then $f(x)$ is the inverse of $f^{-1}(y)$.*

Proof
By definition

$$f^{-1}(f(x)) = x. \tag{4.18}$$

Let's now apply function $f(x)$ to both sides of this equation (this is possible because the range of $f^{-1}(y)$ is the same as the domain of $f(x)$):

$$f\left(f^{-1}(f(x))\right) = f(x). \tag{4.19}$$

Next, we introduce a new notation:

$$y = f(x). \tag{4.20}$$

This is just a notation that defines a new variable y. This variable can have any value, as long as it is in the range of $f(x)$. Using this notation, equation (4.19) is in the form:

$$f\left(f^{-1}(y)\right) = y. \tag{4.21}$$

Since y is in the range of $f(x)$, it is also in the domain of f^{-1}. Then equation (4.21) means that $f(x)$ is the inverse of $f^{-1}(y)$.

For example, function $f(x) = x - 2$ is the inverse for $g(y) = y + 2$. Of course, $g(y) = y + 2$ is the inverse to $f(x) = x - 2$.

4.5.2 Inverse of a Composite Function

Suppose we have a composite function

$$h(x) = g(f(x)). \tag{4.22}$$

Suppose that $g^{-1}(y)$, and $f^{-1}(y)$ exist. We also assume that all domains and ranges match as required. Does $h^{-1}(y)$ exist, and if yes, how do we compute it? The idea is to invert the composite function in steps, starting from the last one.

Theorem 4.6 *The inverse of a composite function $h(y) = g(f(x))$ is given by*

$$h^{-1}(y) = f^{-1}\left(g^{-1}(y)\right). \tag{4.23}$$

Proof

Consider the function

$$H(y) = f^{-1}\left(g^{-1}(y)\right), \tag{4.24}$$

whose domain is the same as the domain of g^{-1} or, equivalently, the range of g. Note that at this point we do not state that it is the inverse of $h(y)$. Let's apply this function to $h(x)$ and see what happens.

$$H(h(x)) = f^{-1}\left(g^{-1}(h(x))\right). \tag{4.25}$$

We substitute $h(x) = g(f(x))$ here:

$$H(h(x)) = f^{-1}\left(g^{-1}(g(f(x)))\right). \tag{4.26}$$

For brevity we denote $f(x) = z$:

$$H(h(x)) = f^{-1}\left(g^{-1}(g(z))\right). \tag{4.27}$$

By definition of the inverse function, $g^{-1}(g(z)) = z$. This yields:

$$H(h(x)) = f^{-1}(z). \tag{4.28}$$

Now, we recall that $z = f(x)$:

$$H(h(x)) = f^{-1}(f(x)). \tag{4.29}$$

By definition of the inverse function, $f^{-1}(f(x)) = x$. This yields:

$$H(h(x)) = x. \tag{4.30}$$

We see that $H(y)$ is the inverse of $h(x)$. We recall that $H(y) = f^{-1}\left(g^{-1}(y)\right)$, which completes the proof.

If a function $h(x)$ can be viewed as a composite function, theorem 4.6 gives us a great tool: the inverse is computed by applying inverses of the constituent functions in the reverse order. It partitions a more complicated problem into two or more simpler ones, which is always a good strategy. We will make good use of this trick in section 5.6.

4.6 Applying a Function to Both Sides of an Inequality

We have applied functions to both sides of equations multiple times in the previous sections, and this always leaves the equation valid (of course, we must conform to domain restrictions when we apply a function to an equation). Applying a function to both sides of an inequality is a bit more tricky. If the function is strictly monotonic, we have:

1. If $a > b$ and function $f(x)$ is strictly monotonically increasing, then $f(a) > f(b)$.
2. If $a > b$ and function $f(x)$ is strictly monotonically decreasing, then $f(a) < f(b)$.

These rules directly follow from the very definition of a monotonic function. For example, since the cubic function is strictly monotonically increasing, we can cube both sides of equation $x^2 > q$ to get $x^6 > q^3$.

Consider the application of using inverse functions on p. 87. The requirement for the cellular network to provide phone service is

$$S(x) \geq S_0, \tag{4.31}$$

where $S(x)$ is the signal strength as a function of the distance x from the cell tower, and S_0 is the minimum signal strength that is required for the phone to work. We also know that function $S(x)$ is strictly monotonically decreasing because the signal attenuates as it propagates farther from the tower. Then the inverse of that function is also decreasing (p. 90). Application of the inverse function S^{-1} will flip the sign of inequality in (4.31):

$$x \leq S^{-1}(S_0). \tag{4.32}$$

This means that the cell tower will provide a service to any phone that is within some fixed distance $S^{-1}(S_0)$. This is important: as long as cell towers are spaced no farther than some maximum distance given by equation (4.32), there will be no holes in the coverage! On 87 we made the same conclusion using a qualitative, nonrigorous argument. Here, we supported that argument mathematically.

If a function is not strictly monotonic, we may have to split its domain into intervals, where it is strictly monotonic in each interval, and consider these intervals separately. For example, consider the function $f(x) = 1/x$, which is monotonically decreasing for $x > 0$ and, separately, for $x < 0$. However, this function is not monotonically decreasing in general because $f(1) > f(-1)$. Consider inequality $y^2 + 1 > y$, which is valid for all values of y. What happens if we apply function $f(x)$ to both sides of this inequality? If $y > 0$, the inequality flips its sign because $f(x)$ is strictly monotonically decreasing for $x > 0$:

$$\frac{1}{y^2+1} < \frac{1}{y}, \tag{4.33}$$

where it is assumed that $y > 0$.

If $y < 0$, we still have $y^2 + 1 > y$, but we no longer can use the assumption that function $f(x)$ is monotonically decreasing. Instead, we obtain:

$$\frac{1}{y^2+1} > \frac{1}{y}. \tag{4.34}$$

Indeed, the left-hand side is positive and must be greater than the negative right-hand side.

The warning about range and domain constraints when applying a function to an equation is also important for applying a function to inequalities.

4.7 Concave and Convex Functions

A function is called *convex* in an interval if for any two points in that interval the plot of the function lies at or below the line that connects these two points (figure 4.3). Of course, this explanation cannot be used in place of a definition and must be formalized.

Consider two points, x_1 and x_2. Next, consider a point in between that is given by $ax_1 + (1-a)x_2$, where $0 < a < 1$. If $a = 1/2$, the new point is exactly in the middle of the interval (x_1, x_2), and if a is close to 0 or 1, the new point will correspondingly be close to x_2 or x_1. This means that as a varies from 1 to 0, the value of $ax_1 + (1-a)x_2$ varies from x_1 to x_2.

Let us consider the relative position of points P and Q in figure 4.3. The vertical position of point Q is just the value of the function at the point $x = ax_1 + (1-a)x_2$:

$$Q = f(ax_1 + (1-a)x_2). \tag{4.35}$$

The vertical position of point P is given by[5]

$$P = af(x_1) + (1-a)f(x_2). \tag{4.36}$$

A convex function has $P \geq Q$ for any x_1, x_2 such that $x_1 \neq x_2$ and for any $0 < a < 1$. If $P > Q$, the function is called *strictly convex*. A function for which $P \leq Q$ is called concave, and if $P < Q$, the function is called strictly concave. In this section, we will only consider strictly convex and strictly concave functions. For brevity, we will drop the "strictly" qualifier and simply refer to these functions as convex or concave.

On p. 66 we talked about optimization as an important application of mathematics. Optimization is usually formulated as finding a minimum (or a maximum) of a function. In practice, a function may have multiple minima, and we must not only find them, but also select the one that works best for the application in hand.

Fortunately, the problem becomes much simpler if the function is convex and its domain satisfies some basic constraints. Then it has only one minimum. Moreover, computer algorithms for finding this minimum are much faster and more robust that in the general case of nonconvex functions.

[5] A proof of this fact is the subject of exercise 17 for this chapter.

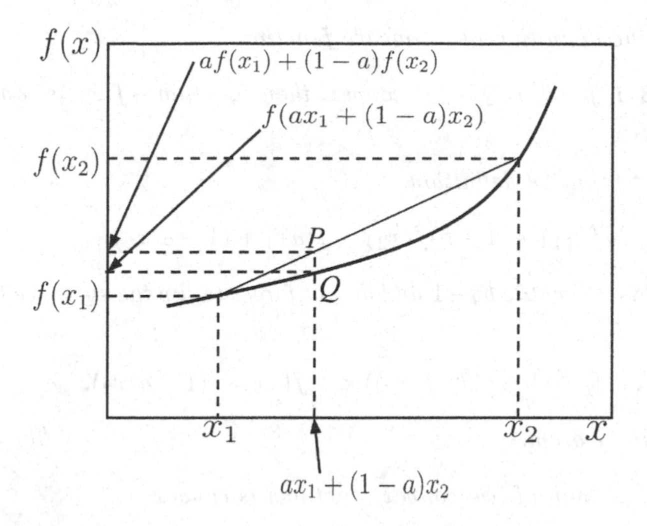

FIGURE 4.3
A convex function

Now, we are ready to prove several theorems about convex and concave functions.

Theorem 4.7 *The inverse of a convex, strictly monotonically increasing function is concave.*

Proof
In the proof, we must somehow use the fact that the function in question is convex. The only way to do that is to use the definition:

$$af(x_1) + (1-a)f(x_2) > f(ax_1 + (1-a)x_2). \tag{4.37}$$

Let's try to apply the inverse function to this inequality. Since function $f(x)$ is monotonically increasing, its inverse is also monotonically increasing (see theorem 4.4). According to the rules on p. 92, the sign of the inequality is then preserved:

$$\begin{aligned} f^{-1}\left(af(x_1) + (1-a)f(x_2)\right) &> f^{-1}\left(f(ax_1 + (1-a)x_2)\right) \\ &= ax_1 + (1-a)x_2. \end{aligned} \tag{4.38}$$

Next, we make new notations: $y_1 = f(x_1), y_2 = f(x_2)$. We also observe that x_1, x_2 in the right-hand side can be written as

$$\begin{aligned} x_1 &= f^{-1}\left(f(x_1)\right) = f^{-1}(y_1), \\ x_2 &= f^{-1}\left(f(x_2)\right) = f^{-1}(y_2). \end{aligned} \tag{4.39}$$

Then inequality (4.38) is as follows:

$$f^{-1}\left(ay_1 + (1-a)y_2\right) > af^{-1}(y_1) + (1-a)f^{-1}(y_2). \tag{4.40}$$

This satisfies the definition of a concave function.

Theorem 4.8 *If function $f(x)$ is convex, then function $-f(x)$ is concave.*

Proof
We again start from the definition:

$$af(x_1) + (1-a)f(x_2) > f(ax_1 + (1-a)x_2). \tag{4.41}$$

We multiply this inequality by -1 and do not forget to flip the sign (see theorem 3.1):

$$a(-f(x_1)) + (1-a)(-f(x_2)) < -f(ax_1 + (1-a)x_2). \tag{4.42}$$

This proves the theorem.

Theorem 4.9 *A sum of two convex functions is convex.*

Proof
Consider two convex functions:

$$\begin{aligned} af(x_1) + (1-a)f(x_2) &> f(ax_1 + (1-a)x_2), \\ ag(x_1) + (1-a)g(x_2) &> g(ax_1 + (1-a)x_2). \end{aligned} \tag{4.43}$$

We add these two inequalities (see theorem 3.2) and collect the terms to get

$$\begin{aligned} &a(f(x_1) + g(x_1)) + (1-a)(f(x_2) + g(x_2)) > \\ &f(ax_1 + (1-a)x_2) + g(ax_1 + (1-a)x_2). \end{aligned} \tag{4.44}$$

This proves the theorem.

Theorem 4.10 *The quadratic function $f(x) = x^2$ is convex.*

The proof uses the following

Lemma 4.1 *For any x_1, x_2, a and $b = 1-a$ it is the case that*

$$(ax_1 + bx_2)^2 = ax_1^2 + bx_2^2 - ab(x_1 - x_2)^2. \tag{4.45}$$

The proof of this lemma is the subject of exercise 27 for this chapter.

Proof
Now theorem 4.10 directly follows from this lemma (recall that $b = 1-a$ and $0 < a < 1$):

$$\begin{aligned} (ax_1 + (1-a)x_2)^2 &= ax_1^2 + (1-a)x_2^2 - a(1-a)(x_1 - x_2)^2 \\ &< ax_1^2 + (1-a)x_2^2. \end{aligned} \tag{4.46}$$

Jensen's inequality extends the convexivity condition to more than two values of x:

Theorem 4.11 *For a convex function $f(x)$ and positive numbers q_n such that*

$$\sum_{k=1}^{n} q_k = 1, \tag{4.47}$$

it is the case that

$$f(q_1x_1 + q_2x_2 + \ldots + q_nx_n) \le q_1f(x_1) + q_2f(x_2) + \ldots + q_nf(x_n). \tag{4.48}$$

Proof

The proof is by induction.

1. *For $n = 2$ the Jensen inequality reduces to the convexivity condition for function $f(x)$. It holds.*
2. *For $n + 1$ we need to prove that*

$$\begin{aligned} &f(q_1x_1 + q_2x_2 + \ldots + q_nx_n + q_{n+1}x_{n+1}) \le \\ &q_1f(x_1) + q_2f(x_2) + \ldots + q_nf(x_n) + q_{n+1}f(x_{n+1}). \end{aligned} \tag{4.49}$$

We multiply and divide by $q_n + q_{n+1}$ the two last terms in the sum of q_nx_n to get:

$$q_nx_n + q_{n+1}x_{n+1} = (q_n + q_{n+1}) \cdot \left(\frac{q_nx_n}{q_n + q_{n+1}} + \frac{q_{n+1}x_{n+1}}{q_n + q_{n+1}} \right). \tag{4.50}$$

Now the sum in the left-hand side of equation (4.49) can be viewed as having just n terms:

$$\begin{aligned} &f(q_1x_1 + q_2x_2 + \ldots + q_{n+1}x_{n+1}) = \\ &f(q_1x_1 + q_2x_2 + \ldots + q_{n-1}x_{n-1} + q'_nx'_n), \end{aligned} \tag{4.51}$$

where we denoted

$$\begin{aligned} q'_n &= q_n + q_{n+1}, \\ x'_n &= \frac{q_nx_n}{q_n + q_{n+1}} + \frac{q_{n+1}x_{n+1}}{q_n + q_{n+1}}. \end{aligned} \tag{4.52}$$

Importantly, we have $q_1 + q_2 + \ldots + q'_n = q_1 + q_2 + \ldots + (q_n + q_{n+1}) = 1$. This means that we can use the Jensen inequality for n terms:

$$\begin{aligned} &f(q_1x_1 + q_2x_2 + \ldots + q_nx_n + q_{n+1}x_{n+1}) = \\ &f(q_1x_1 + q_2x_2 + \ldots + q_{n-1}x_{n-1} + q'_nx'_n) \le \\ &q_1f(x_1) + q_2f(x_2) + \ldots + q_{n-1}f(x_{n-1}) + q'_nf(x'_n) = \\ &q_1f(x_1) + q_2f(x_2) + \ldots + q'_nf\left(\frac{q_nx_n}{q_n + q_{n+1}} + \frac{q_{n+1}x_{n+1}}{q_n + q_{n+1}} \right). \end{aligned} \tag{4.53}$$

The right-hand side contains function $f(x)$ of a sum of two variables where we can again use the convexivity of $f(x)$. This produces the desired Jensen inequality for $n + 1$ values of q, x.

Under certain conditions, combining convex functions can produce a convex function. The rules for that are given by the following theorems.

Theorem 4.12 *If function $f(x)$ is convex and function $g(y)$ is convex and strictly monotonically increasing, then function $g(f(x))$ is convex.*

Proof
Since $f(x)$ is convex, we have

$$af(x_1) + (1-a)f(x_2) > f(ax_1 + (1-a)x_2). \tag{4.54}$$

Let us apply function $g(y)$ to both sides of this inequality. Since $g(y)$ is increasing, the sign of the inequality is preserved (*see p. 92*)*:*

$$g\left(af(x_1) + (1-a)f(x_2)\right) > g\left(f(ax_1 + (1-a)x_2)\right). \tag{4.55}$$

Separately, we use the fact that $g(y)$ is convex:

$$ag(y_1) + (1-a)g(y_2) > g(ay_1 + (1-a)y_2). \tag{4.56}$$

Since this inequality is valid for any values of y_1, y_2, we can apply it to $y_1 = f(x_1), y_2 = f(x_2)$:

$$ag(f(x_1)) + (1-a)g(f(x_2)) > g(af(x_1) + (1-a)f(x_2)). \tag{4.57}$$

From inequalities (4.55) and (4.57) we obtain

$$ag(f(x_1)) + (1-a)g(f(x_2)) > g\left(f(ax_1 + (1-a)x_2)\right). \tag{4.58}$$

This proves the theorem.

Theorem 4.13 *If nonnegative functions $f(x)$ and $g(x)$ are convex and both are strictly monotonically increasing or monotonically decreasing, then function $f(x) \cdot g(x)$ is convex.*

Proof
Consider the definitions for the convexivity of $f(x)$ and $g(x)$:

$$\begin{aligned} af(x_1) + (1-a)f(x_2) &> f(ax_1 + (1-a)x_2), \\ ag(x_1) + (1-a)g(x_2) &> g(ax_1 + (1-a)x_2). \end{aligned} \tag{4.59}$$

Since the left and right-hand sides of these inequalities are nonnegative, we can multiply them. We expand the parentheses in the resulting inequality to get:

$$\begin{aligned} &a^2 f(x_1)g(x_1) + (1-a)^2 f(x_2)g(x_2) + \\ &a(1-a)f(x_1)g(x_2) + a(1-a)f(x_2)g(x_1) > \\ &f(ax_1 + (1-a)x_2)g(ax_1 + (1-a)x_2). \end{aligned} \tag{4.60}$$

Our goal is to get the inequality for the convexivity of function $f(x)g(x)$. We expect it to look as follows

$$\begin{aligned} &af(x_1)g(x_1) + (1-a)f(x_2)g(x_2) > \\ &f(ax_1 + (1-a)x_2)g(ax_1 + (1-a)x_2). \end{aligned} \tag{4.61}$$

In inequality (4.60) we already got the right-hand side of inequality (4.61), but the left-hand sides of these two inequalities differ. We add and subtract the terms that we expect to have in the left-hand side for convexivity, but which are missing in (4.60):

$$\begin{aligned} &af(x_1)g(x_1) + (1-a)f(x_2)g(x_2) - \\ &af(x_1)g(x_1) - (1-a)f(x_2)g(x_2) + \\ &a^2 f(x_1)g(x_1) + (1-a)^2 f(x_2)g(x_2) + \\ &a(1-a)f(x_1)g(x_2) + a(1-a)f(x_2)g(x_1) > \\ &f(ax_1 + (1-a)x_2)g(ax_1 + (1-a)x_2). \end{aligned} \tag{4.62}$$

We introduce new notations:

$$\begin{aligned} E &= -af(x_1)g(x_1) - (1-a)f(x_2)g(x_2) \\ &\quad + a^2 f(x_1)g(x_1) + (1-a)^2 f(x_2)g(x_2) \\ &\quad + a(1-a)f(x_1)g(x_2) + a(1-a)f(x_2)g(x_1), \\ P &= af(x_1)g(x_1) + (1-a)f(x_2)g(x_2), \\ Q &= f(ax_1 + (1-a)x_2)g(ax_1 + (1-a)x_2). \end{aligned} \tag{4.63}$$

With these notations, the convexivity condition (4.61) of function $f \cdot g$ is simply $P > Q$. Instead of $P > Q$, we so far have inequality (4.62), which is written in new notations as

$$P + E > Q. \tag{4.64}$$

Our goal is to show that inequality (4.64) implies the desired inequality $P > Q$. To do that, we should look at the "extra" term E. We collect terms in the right-hand side of the definition of E in equations (4.63):

$$\begin{aligned} E = &- a(1-a)f(x_1)g(x_1) - a(1-a)f(x_2)g(x_2) \\ &+ a(1-a)f(x_1)g(x_2) + a(1-a)f(x_2)g(x_1). \end{aligned} \tag{4.65}$$

The factor $a(1-a)$ is common to all terms:

$$E = a(1-a)(-f(x_1)g(x_1) - f(x_2)g(x_2) + f(x_1)g(x_2) + f(x_2)g(x_1)). \tag{4.66}$$

The sum containing f and g can be factored:

$$E = a(1-a)(f(x_1) - f(x_2))(g(x_2) - g(x_1)). \tag{4.67}$$

If both functions f and g are monotonically increasing or monotonically decreasing, then factors $(f(x_1)-f(x_2))$ and $(g(x_2)-g(x_1))$ have different signs. Then, we have

$$E < 0. \tag{4.68}$$

Now, we turn back to inequality (4.64), which we already established. We see that

$$P > Q - E > Q. \tag{4.69}$$

This proves the theorem.

Exercises

1. We know that any function with a symmetric domain is a sum of an even and an odd function. Suppose $f = f_e + f_o$ and $g = g_e + g_o$, where subscripts o and e denote an odd and an even function. Define $h = fg$. This function in turn can be represented as $h = h_e + h_o$. Express h_e, h_o through f_e, f_o, g_e, g_o. Do the same for f/g. (Hint: use expressions for the product and ratio of complex numbers as a template.) What does this say when f and g are both even, both odd, or one is even and another is odd?
2. Prove that x^3 is monotonic. (Hint: use $a^3-b^3 = (a-b)(a^2+ab+b^2)$. The second factor in the right-hand side was a subject of exercise 6 for chapter 3.)
3. For even and odd functions consider $f_e(f_o(x)), f_o(f_e(x)), f_e(f_e(x))$, and $f_o(f_o(x))$. Which ones are even and which are odd?
4. Give another example of a function that is not monotonic, but has an inverse.
5. Prove that function x^{2n} is convex.
6. Where in the proof of theorem 4.13 did we use $0 < a < 1$?
7. For all real x, we compute a random value in the interval from 0 to 1. Is this a function?
8. Give an example of the weirdest function that you can imagine. The function should be defined on a subset of real numbers. Define the domain, the range, and the association between the two.
9. Is there a function that is both even and odd? If yes and if its domain is all real numbers, is it unique?
10. Take function $f(x) = x^2 + x^3$ and construct the even and odd components $f_e(x)$ and $f_o(x)$ using theorem 4.1. After constructing

$f_e(x)$ and $f_o(x)$, reconstruct back the original function $f(x) = x^2 + x^3$. Does it work?

11. What will we get for the even and odd components if $f(x)$ is a polynomial: $f(x) = a_n x^n + a_{n-1}x^{n-1} + ... + a_1 x + a_0$?
12. Prove that if a function is strictly monotonically decreasing, then its inverse is also strictly monotonically decreasing.
13. Prove that the sum of two monotonically increasing functions is also monotonically increasing.
14. Prove that if an even function is monotonically increasing for $x > 0$, it is monotonically decreasing for $x < 0$.
15. Prove that if an odd function is monotonically increasing for $x > 0$, it is monotonically increasing for $x < 0$.
16. A person in front of the White House holds a sign: "A product of two monotonic functions is monotonic!" Prove him wrong. What conditions must satisfy monotonically increasing functions $f(x)$ and $g(x)$ in order for function $f(x) \cdot g(x)$ to be monotonically increasing?
17. Refer to figure 4.3. Prove that points $(x_1, f(x_1)), (x_2, f(x_2))$, and $(ax_1 + (1-a)x_2, af(x_1) + (1-a)f(x_2))$ lie on a straight line.
18. Prove that the inverse of a convex, strictly monotonically decreasing function is convex.
19. Prove that functions $f(x) = \sqrt{x}$ and $g(x) = \sqrt[4]{x}$ are concave. (Both functions are defined for $x \geq 0$.)
20. Prove that a product of two convex functions is not necessarily convex.
21. Using theorem 4.13 as a template, prove that if a nonnegative function $f(x)$ is defined for $x \geq 0$ and is strictly convex, then function $xf(x)$ is strictly convex.
22. Prove that function x^n is convex for $x \geq 0, n \geq 2$.
23. Is it true that an odd function that is convex for $x \geq 0$ is concave for $x < 0$? Is it true that an even function that is convex for $x \geq 0$ is convex for all values of x?
24. For what values of parameter p is function $f(x) = \sqrt{|x| - p}/(x^2 + p^2)$ defined for all x? Answer the same question for function $f(x) = \sqrt{|x| - p}/(x^2 + p^2 + 1)$.
25. Prove that if $f(g(x)) = g(f(x))$, then the inverse functions also commute.
26. Prove that the difference of a convex and a concave function is convex.
27. Prove the lemma for theorem 4.10 (p. 96).

28. In the proof of the Jensen inequality (theorem 4.11), we conveniently omitted the requirement that x_n' must be in the interval where function $f(x)$ is convex. Prove that this is indeed the case.
29. For the function that is defined by equation (4.14) we stated that $w(x)$ is irrational iff x is irrational. Prove that this is the case.
30. Use convexity of function $f(x) = x^2$ to prove inequality (3.30) for the arithmetic and quadratic mean.

5

Polynomials

5.1 Definition and Examples

A polynomial is an algebraic expression containing one or more variables and *coefficients*, where the variables are combined with coefficients using only addition, subtraction, multiplication, and exponentiation with nonnegative integer powers. In this name, "poly" means "many" and refers to possibly having many additive terms with different powers of the variable or variables. A polynomial may also be called a *monomial* if it has just one term, or a *binomial* if it has two terms.

In this chapter, we will only study polynomials of a single variable and binomials. Let's denote the variable in a polynomial as x. According to our definition, the following expressions are polynomials:

1. A binomial $P_1(x) = 2x + 5$. In this expression, we have the sum of two terms, $2x$ and 5. In the first term, we have a product of two factors, 2 and x, where 2 is a coefficient and x is the variable.
2. $P_2(x) = ax^2 + bx + c$, where a, b and c are coefficients. The variable x is raised to the second and first power; if $x \neq 0$, the last term c may be viewed as having x raised to power 0.
3. $P_3(x) = ax^2 + bx + c + 2 + 3x + 4x^2$. It is a sum of six terms, each formed by multiplication and a nonnegative integer power of x. It is customary to collect terms by powers of x in an expression like this to have this polynomial in the *canonical form*: $P(x) = (a+4)x^2 + (b+3)x + (c+2)$.
4. A monomial $P_4(x) = 3x$ is also a polynomial, albeit a simple one.

Many theorems in this chapter will apply to polynomials with any number of terms, and we will widely use the sigma notation that we introduced on p. 75. A general expression for a polynomial is as follows:

$$\sum_{n=0}^{N} a_n x^n = a_0 + a_1 x + a_2 x^2 + ... + a_{N-1} x^{N-1} + a_N x^N. \tag{5.1}$$

DOI: 10.1201/9781003456766-5

Consider, for example, a quadratic polynomial $P_2(x) = \pi + 3x + 2x^2$. Since the highest power in this polynomial is 2, we get $N = 2$:

$$P_2(x) = \sum_{n=0}^{2} a_n x^n = a_0 + a_1 x^1 + a_2 x^2, \tag{5.2}$$

where $a_0 = \pi, a_1 = 3$, and $a_2 = 2$.

5.2 Binomial Expansion

We have already introduced a binomial as a mathematical expression with two additive terms. For example, $a + b$ or $cx + d$ are binomials. Let's see what happens if we raise a binomial $a + b$ to the n-th power and expand the result. We will have to deal with the following expression that contains n factors:

$$(a+b)^n = \underbrace{(a+b) \cdot (a+b) \cdot \ldots \cdot (a+b)}_{n}. \tag{5.3}$$

When expanding this product, we have to pull either a or b from each pair of parentheses. Each additive term in the resulting expression will have some number of a and some number of b factors, but the *total* number of a and b factors in each term will be n because there are n pairs of parentheses, and either a or b is pulled from each pair. Therefore, all additive terms will have the form $a^m b^k$, such that $m + k = n$. A more explicit and concise expression is $a^k b^{n-k}$, where we automatically ensure that the total number of a and b factors amounts to n.

For example, consider a term that is formed when we pull just a from each pair of parentheses. This will produce the term a^n in the result and will correspond to $a^k b^{n-k}$, where $k = n$. In another example, we pull just b from each pair of parentheses; this will produce b^n and will correspond to $k = 0$.

Next, consider what happens if we pull a from each pair of parentheses, except the first one, from which we pull b. This will produce the term $a^{n-1}b$, which corresponds to $k = n - 1$. The same will apply to the case, when b is pulled from the second pair of parentheses, the third, or the n-th. We conclude that there are n different ways to produce the term $a^{n-1}b$. If we collect these terms, the result will be $na^{n-1}b$. This is different from the cases of a^n and b^n, each produced by only one selection of a and b.

We see that the whole expression (called the *binomial expansion*) is given by

$$(a+b)^n = \sum_{k=0}^{n} \binom{n}{k} a^k b^{n-k}, \tag{5.4}$$

where $\binom{n}{k}$ are some numerical coefficients, yet to be determined (this notation is specific to binomial coefficients).

Consider how the binomial coefficient $\binom{n}{k}$ is formed. It is the number of occurrences of terms $a^k b^{n-k}$ in the full expansion of the right-hand side in equation (5.3). Each such term is formed by selecting k such $(a+b)$ factors in the right-hand side of equation (5.3) that contribute an a to the product $a^k b^{n-k}$. (The remaining $n-k$ factors automatically contribute a b each). Each term $a^k b^{n-k}$ in the expansion adds one count to the binomial coefficient $\binom{n}{k}$ in the right-hand side of equation (5.4). It is simply the number of ways to select k factors from a set of n factors.

Now let us consider a general problem of selecting k objects from a set of n objects, such as pencils on a table, or stars in the sky. It is clear that the number of ways to do that must not depend on the nature of the objects we are selecting, but must depend only on the values of n and k. Then it is the same as the number of ways to select k factors in the right-hand side of equation (5.3)—that is, equal to the binomial coefficient $\binom{n}{k}$.

Since the probability of randomly selecting a particular combination of objects depends on the number of possible selections, binomial coefficients frequently appear in probability applications. For example, the chances of winning a game of chance (including the lottery) is often computed using binomial coefficients. In another example, consider a company that operates a fleet of 100 delivery trucks. To ensure a smooth operation, they need at least 95 trucks to be on the road, with up to 5 trucks being repaired or serviced. What are the chances that on a particular day 6 trucks break down, jeopardizing the business of that company? The answer to this very real business question involves computing the binomial coefficient $\binom{100}{6}$.

So far, we have learned that $\binom{n}{0} = 1$, $\binom{n}{n} = 1$, and $\binom{n}{1} = n$. What about all the other coefficients? The following theorem gives a convenient way to compute binomial coefficients $\binom{n}{k}$ for any n, k:

Theorem 5.1 *Binomial coefficients can be computed recursively using the following formula:*[1]

$$\binom{n+1}{k} = \begin{cases} 1, & \text{if } k = n+1 \text{ or } k = 0, \text{ and} \\ \binom{n}{k-1} + \binom{n}{k} & \text{if } 0 < k \leq n. \end{cases} \tag{5.5}$$

[1]Recursion is a method to compute a result for some parameter n using one or more values for smaller values of that parameter.

Proof

We have already established the first case ($k = 0$ or $k = n+1$) in (5.5). Here, we prove the second case. The proof is by induction.

1. *For $n = 2$ we have*

$$(a+b)^2 = a^2 + 2ab + b^2, \tag{5.6}$$

which implies that $\binom{2}{1} = 2$. Indeed, the second case in (5.5) yields:

$$\binom{2}{1} = \binom{1}{0} + \binom{1}{1} = 1 + 1. \tag{5.7}$$

2. *We assume that the theorem is true for n. For power $(n+1)$ we get:*

$$(a+b)^{n+1} = \left(\sum_{m=0}^{n} \binom{n}{m} a^m b^{n-m} \right) \cdot (a+b). \tag{5.8}$$

Let's consider the coefficient for the term that contains a^k if we expand the right-hand side. There are only two ways to get a term with a^k term in the expansion:

(a) *We pull the term with a^k from the factor that contains the sum with the binomial coefficients, and pull the b from the factor $(a+b)$. The term with a^k in the sum contains $\binom{n}{k}$, and this value contributes to the coefficient for a^k in the final result.*

(b) *We pull the term with a^{k-1} from the factor that contains the sum with the binomial coefficients, and pull the a from the factor $(a+b)$. This contributes the value of $\binom{n}{k-1}$ to the coefficient for the term containing a^k in the final result.*

In both cases, the power of b will be $n+1-k$. The resulting coefficient for $a^k b^{n+1-k}$ is then given by $\binom{n}{k-1} + \binom{n}{k}$.

This proves the theorem.

This theorem expresses any coefficient $\binom{n+1}{k}$ for a binomial of degree $n+1$ through the coefficients for a binomial of a lower degree n. If we apply this theorem successively multiple times, there will appear binomial coefficients $\binom{i}{j}$, where $i < n+1$ and $j \leq i$. Some of these binomial coefficients will have $i = j$ or $j = 0$; they will be equal to 1. For all the other ones we will continue to apply theorem 5.1 until we get to $\binom{1}{1}$ and $\binom{1}{0}$, which both are equal to 1. This provides a method to compute the value of a binomial coefficient for any degree.

Using this recursive computation as a starting point, we obtain an explicit formula for binomial coefficients, which yields a direct way to compute them:

Theorem 5.2 *Binomial coefficients* $\binom{n}{k}$ *are given by:*

$$\binom{n}{k} = \frac{n!}{k!(n-k)!}. \tag{5.9}$$

Here, the notation $n!$ is called "n factorial" and means $n! = 1 \cdot 2 \cdot 3 \cdot \ldots \cdot n$. By convention, we use $0! = 1$.

Proof

The proof is by induction.[2]

1. *For* $n = 1$ *we have*

$$(a+b)^1 = a + b, \tag{5.10}$$

which corresponds to $\binom{1}{1} = \binom{1}{0} = 1$. *A check of equation (5.9) shows that the theorem is satisfied for the case* $n = 1$.

2. *We assume that the theorem is true for power* n:

$$(a+b)^n = \sum_{k=0}^{n} \binom{n}{k} a^k b^{n-k}, \tag{5.11}$$

where

$$\binom{n}{k} = \frac{n!}{k!(n-k)!}. \tag{5.12}$$

We use theorem 5.1 and substitute the explicit expressions for the binomial coefficients to get a formula for $n+1$:

$$\begin{aligned} \binom{n+1}{k} &= \binom{n}{k} + \binom{n}{k-1} \\ &= \frac{n!}{k!(n-k)!} + \frac{n!}{(k-1)!(n-(k-1))!}. \end{aligned} \tag{5.13}$$

Our goal is to get a common denominator for these two fractions. To do that, we multiply both the numerator and the denominator by $n+1-k$ *for the first term and by* k *for the second term:*

$$\begin{aligned} \binom{n+1}{k} &= \binom{n}{k} + \binom{n}{k-1} \\ &= \frac{n!(n+1-k)}{k!(n-k)!(n+1-k)} + \frac{n!k}{(k-1)!k(n+1-k)!}, \end{aligned} \tag{5.14}$$

where we used $n-(k-1) = n+1-k$. *We observe that in the denominator of the first term we get*

$$k!(n-k)!(n+1-k) = k!(n+1-k)!. \tag{5.15}$$

[2] Also see exercise 22 for this chapter.

Similarly, in the denominator of the second term we get

$$(k-1)!k(n+1-k)! = k!(n+1-k)!. \tag{5.16}$$

Then both terms have a common denominator. This produces

$$\binom{n+1}{k} = \frac{n!(n+1-k)+n!k}{k!(n+1-k)!}. \tag{5.17}$$

In the numerator, we factor out $n!$*; this simplifies the numerator to* $n!(n+1)$*, which is equal to* $(n+1)!$*. The final expression is*

$$\binom{n+1}{k} = \frac{(n+1)!}{k!((n+1)-k)!}, \tag{5.18}$$

which proves the theorem.

If we substitute the explicit expressions for the factorials in equation (5.9), we will get an alternative form for the binomial coefficients. We start from

$$\begin{aligned}\binom{n}{k} &= \frac{n!}{k!(n-k)!}\\ &= \frac{1\cdot 2\cdot 3\cdot\ldots\cdot n}{k!\cdot 1\cdot 2\cdot 3\cdot\ldots\cdot(n-k)}.\end{aligned} \tag{5.19}$$

The first $n-k$ factors in the numerator and the same factors in the denominator cancel. We get

$$\binom{n}{k} = \frac{(n-k+1)\cdot\ldots\cdot n}{k!}. \tag{5.20}$$

We will use this expression in section 9.12.

One more note deals with proving Bernoulli's inequality (we explored this inequality in section 3.3.7):

$$(1+p)^n \geq 1+pn. \tag{5.21}$$

Here, we prove it for any $p \geq 0$ and for a natural n. Suppose that we expand the left-hand side using a binomial expansion. Then the first two terms in the binomial expansion will match the right-hand side $1+pn$, and all the additional terms of the expansion are nonnegative for $p \geq 0$. This proves Bernoulli's inequality for nonnegative p. Compare this with a stronger theorem 3.18, which extends the condition for p and requires only that $p > -1$.

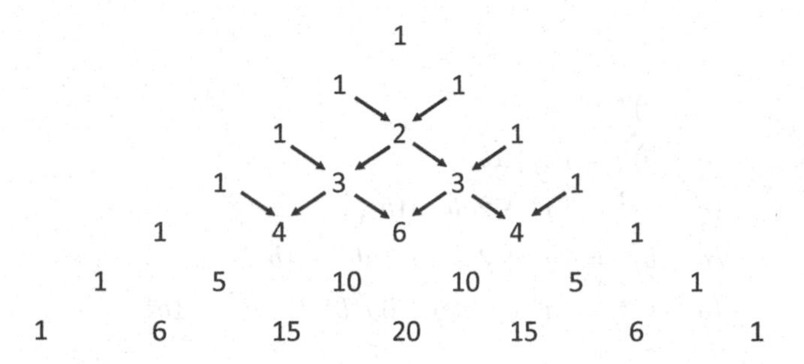

FIGURE 5.1
Pascal's triangle

5.3 Pascal's Triangle

Pascal's triangle is shown in figure 5.1. It is an arrangement of numbers, where each line starts and ends with 1 and where all the other entries are formed by summing the two neighboring entries in the previous line. For example, number 3 is formed as a sum of 1 and 2 in the previous line; number 6 is computed as a sum of $3 + 3$ and so on.

> The name of Pascal's triangle suggests that Blaise Pascal (1623 — 1662) was its discoverer, but (unknown to Pascal), Indian scholars wrote about this elegant construction long before that. The earliest idea about this triangle may have been formulated by Pingala in or before the 2nd century BCE.
>
> The binomial coefficients frequently arise when we need to compute the number of possible combinations for arranging some objects (see p. 105). Pingala wrote a treatise on Sanskrit prosody—that is, the patterns of rhythm and sound used in poetry. He tried to determine the possible number of poetic lines of a certain length with different arrangements of sounds, such as different patterns of using long and short vowels. This combinatorics problem naturally leads to binomial coefficients. The solution of this problem as described in ancient Indian manuscripts shows the arrangement of numbers that today would be called Pascal's triangle. Regretfully, the application of Pascal's triangle to poetry is quite rare in our time.

The numbers in Pascal's triangle are binomial coefficients. Indeed, compare the coefficients in the binomial expansions below with the entries in Pascal's triangle:

$$\begin{aligned}
(a+b)^0 &= 1, \\
(a+b)^1 &= 1a + 1b, \\
(a+b)^2 &= 1a^2 + 2ab + 1b^2, \\
(a+b)^3 &= 1a^3 + 3a^2b + 3ab^2 + 1b^3, \\
(a+b)^4 &= 1a^4 + 4a^3b + 6a^2b^2 + 4ab^3 + 1b^4.
\end{aligned} \tag{5.22}$$

The mechanics behind this way of computing binomial coefficients is given by theorem 5.1, which computes a binomial coefficient for power $n+1$ as a sum of two neighboring binomial coefficients for power n. Pascal's triangle is simply a way to display this theorem graphically. It has a number of interesting properties; here we prove only one of them.

Theorem 5.3 *The sum of the binomial coefficients for $(a+b)^n$ (and of the numbers in the n-th row of Pascal's triangle) is equal to 2^n. (Here, we assume that enumeration of rows in Pascal's triangle starts from 0.)*

We give two proofs of this theorem.

Proof

Binomial coefficients are defined by equation (5.4) as

$$(a+b)^n = \sum_{k=0}^{n} \binom{n}{k} a^k b^{n-k}. \tag{5.23}$$

In this equation we set $a = b = 1$ to get

$$2^n = \sum_{k=0}^{n} \binom{n}{k}. \tag{5.24}$$

The second proof is by induction and uses Pascal's triangle.

Proof

1. *For $n = 0$ and $n = 1$ the theorem holds.*
2. *We assume that the theorem holds for n and consider the n-th and $(n+1)$-th rows of Pascal's triangle. We augment each row by adding a zero entry in the beginning and at the end (figure 5.2).*

   In the $(n+1)$-th row each entry is computed as a sum of two entries in the n-th row. Having additional zeroes makes this rule applicable for the first and the last entry as well. For example:

   (a) *The first and the last nonzero entries in the $(n+1)$-th row are equal to 1 and are computed as a sum of 0 and 1 in the n-th row.*

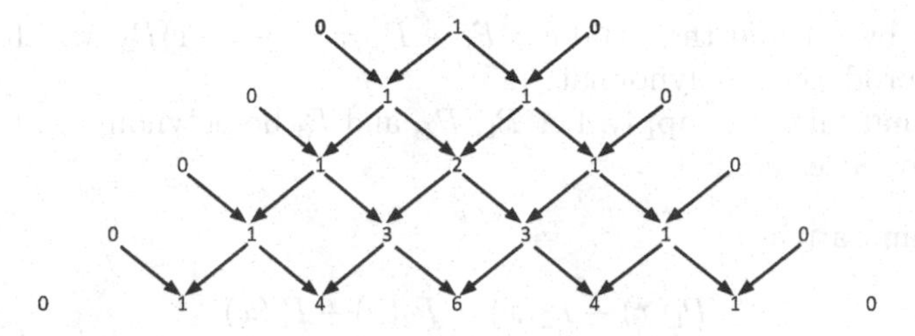

FIGURE 5.2
Pascal's triangle augmented with zeros

(b) *The second nonzero entry in the $(n+1)$-th row is equal to $\binom{n+1}{1}$ and is computed as a sum of 1 and $\binom{n}{1}$ in the n-th row.*

From this we can see that each entry in the n-th row contributes twice to the values of the entries in the $(n+1)$-th row, first to the entry on the left and second to the one on the right. Therefore, the sum of all the values in the $(n+1)$-th row must be equal to the double the sum of the entries in the n-th row. Thus, if the sum of the entries in the n-th row is equal to 2^n, the sum for the $(n+1)$-th row must be equal to 2^{n+1}.

This completes our exploration of binomials. In the following sections, we shift our focus to polynomials proper.

5.4 Adding and Multiplying Polynomials

Polynomials can be added and subtracted according to general algebraic rules. Since addition and subtraction are "allowed" operations in the definition of a polynomial, the result must be a polynomial. For example, consider adding a cubic and a quadratic polynomial:

$$\begin{aligned}&(a_3x^3 + a_2x^2 + a_1x + a_0) + (b_2x^2 + b_1x + b_0) = \\ &a_3x^3 + (a_2 + b_2)x^2 + (a_1 + b_1)x + (a_0 + b_0).\end{aligned} \tag{5.25}$$

The degree of the resulting polynomial is the same or lower than the larger degree of the two polynomials that are being added.

Multiplication is also an "allowed" operation, in the sense that a product of two polynomials is a polynomial. Therefore, the operations of addition and multiplication for polynomials have the property of closure, similar to that property for real numbers. Since a constant can be viewed as a polynomial of zeroth degree, the multiplication of a polynomial by a constant also produces a polynomial. The subtraction of two polynomials is the same as multiplying one of them by -1 and then adding: $P_1 - P_2 = P_1 + (-1)P_2$, which means that it also produces a polynomial.

Other usual rules also apply. Let P_1, P_2, and P_3 be polynomials, then the following properties hold:

1. Commutativity:

$$P_1(x) + P_2(x) = P_2(x) + P_1(x),$$
$$P_1(x) \cdot P_2(x) = P_2(x) \cdot P_1(x). \tag{5.26}$$

2. Associativity:

$$(P_1(x) + P_2(x)) + P_3(x) = P_1(x) + (P_2(x) + P_3(x)),$$
$$(P_1(x) \cdot P_2(x)) \cdot P_3(x) = P_1(x) \cdot (P_2(x) \cdot P_3(x)). \tag{5.27}$$

3. Distributivity:

$$P_3(x) \cdot (P_1(x) + P_2(x)) = P_3(x) \cdot P_1(x) + P_3(x) \cdot P_2(x). \tag{5.28}$$

These properties apply to the expressions for the coefficients of the sums and products of two polynomials. For example, the formulas for the coefficients of $P_1(x) + P_2(x)$ are the same as the formulas for the coefficients of $P_2(x) + P_1(x)$. The commutativity and associativity of addition are the easiest to prove, as they directly follows from the definition. The formulas for the coefficients of a product of two polynomials are a bit more difficult to derive. In turn, these formulas help prove the distributivity property and the commutativity of the product. Even more cumbersome is the proof of the associativity of the product of polynomials.

Polynomials are simpler than many other functions, and mathematicians developed powerful tools to study and manipulate polynomials. As it happens, it is possible to approximate many nonpolynomial functions using polynomials. Instead of dealing with the zoo of complicated mathematical functions, many of them with poorly known properties, scientists use polynomials to build approximate models. In many cases it is a choice between having an approximate polynomial solution and having no solution at all.

5.5 When Are Two Polynomials Equal?

Consider two polynomials, $P_1(x)$ and $P_2(x)$. Let's consider the case when $P_1(x) = P_2(x)$ for any value of x. In mathematics, this condition is referred to as the two polynomials being *identically equal* to each other and is denoted as

$$P_1(x) \equiv P_2(x). \tag{5.29}$$

For this condition we prove the following theorem:

Theorem 5.4 *Two polynomials are identically equal if and only if their coefficients are pairwise equal.*

Consider the explicit expressions for these two polynomials:

$$\begin{aligned} P_1(x) &= a_0 + a_1 x + ... + a_n x^n, \\ P_2(x) &= b_0 + b_1 x + ... + b_m x^m. \end{aligned} \tag{5.30}$$

The theorem states that $P_1(x) \equiv P_2(x)$ iff their degrees are equal ($n = m$) and

$$\begin{aligned} a_0 &= b_0, \\ a_1 &= b_1, \\ &... \\ a_n &= b_n. \end{aligned} \tag{5.31}$$

Proof

1. *We prove that if two polynomials are identically equal, their coefficients are pairwise equal.*

   This statement is proved by induction. To apply induction, we need to enumerate the degrees of the two polynomials. However, here we face a small difficulty: which of the two polynomials' degrees should we use for induction? Note that so far we have not proved that the degrees of these two polynomials are equal (this is a part of the statement of this theorem).

   Without loss of generality we assume that $n \geq m$*. Then, we can use* n *to enumerate the cases in induction. The two steps of induction are as follows:*

   (a) *For* $n = 0$ *we must have* $m = 0$*. Then both polynomials are simply constants, and from* $P_1(x) \equiv P_2(x)$ *it follows* $a_0 = b_0$*.*

   (b) *We assume that the theorem is true for* $n-1$ *and prove it for* n*. Let's use* $x = 0$ *in equation (5.29) and then substitute explicit expressions for the two polynomials there. All the terms with nonzero powers of* x *vanish, and we are left with*

   $$a_0 = b_0. \tag{5.32}$$

   Now, we use $x \neq 0$ *and consider*

   $$a_0 + a_1 x + ... + a_n x^n \equiv b_0 + b_1 x + ... + b_m x^m. \tag{5.33}$$

Since $a_0 = b_0$, we cancel these terms. All the remaining terms have nonzero powers of x. Since $x \neq 0$, we can divide the resulting equation by x to get:

$$a_1 + ... + a_n x^{n-1} \equiv b_1 + ... + b_m x^{m-1}. \quad (5.34)$$

We get two polynomials of lesser degrees equal for all x, except, possibly, for $x = 0$. (The case $x = 0$ for these two polynomials should be treated separately; it is not considered in this book, but it does show that equation (5.34) holds for $x = 0$ as well.) By our induction assumption we see that their coefficients are pairwise equal. Together with the earlier statement $a_0 = b_0$ this proves that if two polynomials are identically equal, then their coefficients are pairwise equal.

2. *The proof of the converse statement (if the coefficients are pairwise equal, then the two polynomials are identically equal) is by a direct substitution.*

5.6 Roots

Consider a polynomial $P(x)$. What are the values of x that produce $P(x) = 0$? This is the same as finding the solutions of the equation $P(x) = 0$. The solutions of a polynomial equation are called *roots*. This term may be used for other types of equations as well, but for polynomials it is used more often.

The Persian mathematician Sharaf al-Dīn al-Ṭūsī (c. 1135 – c. 1213) proved theorems for the number of positive roots of a cubic equation. At that time, mathematicians in the Islamic tradition recognized only positive numbers, so the number of positive roots was perceived as the total number of roots.

General formulas for the roots of a cubic equation were developed by Niccolò Tartaglia and published by Gerolamo Cardano (1501-1576). Today, these formulas bear the name of Cardano, along with the name of the Cardan shaft that is used in many vehicles to transmit the rotary motion to the wheels. Cardano left a fascinating memoir that gives us a peek into the life of a mathematician, medic, and astrologer in the 16th century.

You may be familiar with finding the roots of linear and quadratic equations (both are polynomial). If $a_1 x + a_0 = 0$, then $x = -a_0/a_1$ (assuming $a_1 \neq 0$). A proof of this solution is straightforward: we subtract a_0 and then divide by a_1. For the quadratic equation, the solution is given by the familiar quadratic formula:

Theorem 5.5 *Equation $ax^2 + bx + c = 0$, where $a \neq 0$, has two roots:*

$$x_{1,2} = \frac{-b \pm \sqrt{b^2 - 4ac}}{2a}, \tag{5.35}$$

where $\pm$ denotes two options for the sign, corresponding to the two roots, x_1 and x_2. If the so-called discriminant $b^2 - 4ac$ is nonnegative, the quadratic equation has two real roots. If it is negative, the equation has two distinct complex roots.

Proof

We start by dividing the quadratic equation by a. (Here, we use the condition $a \neq 0$.)

$$x^2 + \frac{b}{a}x + \frac{c}{a} = 0. \tag{5.36}$$

Next, we add and subtract $b^2/(4a^2)$ to the left-hand side:

$$x^2 + \frac{b}{a}x + \frac{b^2}{4a^2} - \frac{b^2}{4a^2} + \frac{c}{a} = 0. \tag{5.37}$$

The first three terms form a complete square:

$$\left(x + \frac{b}{2a}\right)^2 - \frac{b^2}{4a^2} + \frac{c}{a} = 0. \tag{5.38}$$

We move the free terms to the right-hand side and extract the square root:

$$x_{1,2} + \frac{b}{2a} = \pm\sqrt{\frac{b^2}{4a^2} - \frac{c}{a}}. \tag{5.39}$$

Note that the square root as a function of its argument produces only one value (which is called the principal value), but there are two possible values of $x + b/(2a)$ that satisfy equation (5.38).[3] *To capture both possibilities for x, we use the $\pm$ notation for the sign of the square root and the subscripts $1, 2$ for x. We factor out $1/(4a^2)$ from the expression under the square root, move the free term from the left-hand side to the right, and bring both terms to the common denominator:*

$$x_{1,2} = \frac{-b \pm \sqrt{b^2 - 4ac}}{2a}, \tag{5.40}$$

What happens if the discriminant is zero? Then the quadratic formula produces the same value for x_1 and x_2. There still are two roots, but now these roots have the same value. We say that the root multiplicity here is 2.

[3] This is, of course, due to the fact that function $f(y) = y^2$ does not satisfy the horizontal line test and therefore does not have a proper inverse.

This derivation of the quadratic formula hinges on a "lucky guess": we added and subtracted $b^2/(4a^2)$ to equation (5.36). Is there a more intuitive way to obtain the quadratic formula? Here is another derivation. We look at how we solve the linear equation and then try to extend this approach to the quadratic one. Observe that to compute the left-hand side of the linear equation $a_1x + a_0 = 0$ in the beginning of this section, we apply the following operations to the unknown, x:

1. We multiply x by a_1.
2. We add a_0 to the result.

To determine x, we apply the inverse operations in the reverse order (see section 4.5.2). Namely,

1. We subtract a_0.
2. We divide the result by a_1.

This trick is possible because x enters this equation only once, which makes the sequence of steps clear.

For a second-degree polynomial, things get more complicated. Let's first consider an equation that we can solve step by step, like we did for the first-degree polynomial. Here is a quadratic equation, in which x enters the equation only once:

$$p(x - g)^2 - h = 0. \tag{5.41}$$

In essence, this is a particular form of a quadratic equation. Indeed, if we expand the parentheses, we get

$$px^2 - 2pgx + pg^2 - h = 0. \tag{5.42}$$

To compute the left-hand side of equation (5.41) for a given value of x, we need to perform the following operations:

1. Subtract g from x.
2. Square the result.
3. Multiply by p.
4. Subtract h.

Then to compute x for the given values of $p, g,$ and h we use similar tactics as we do for the linear equation: we apply the inverse operations to equation (5.41) in the reverse order:

1. Add h.
2. Divide by p.
3. Compute the square root.
4. Add g.

This leads to the following sequence of equations:

$$p(x-g)^2 = h,$$
$$(x-g)^2 = \frac{h}{p}, \tag{5.43}$$
$$x_{1,2} - g = \pm\sqrt{\frac{h}{p}}.$$

Just as we stated on p. 115, both $x - g = \sqrt{h/p}$ and $x - g = -\sqrt{h/p}$ satisfy equation $(x - g)^2 = h/p$. This is indicated by the $\pm$ notation in the last equation and by subscripts 1,2 for x. Finally, we get

$$x_{1,2} = g \pm \sqrt{\frac{h}{p}}. \tag{5.44}$$

For now, we have learned how to solve a quadratic equation that is written in a special form given by (5.41). However, our goal is to solve a quadratic equation that is written in the canonical form:

$$ax^2 + bx + c = 0. \tag{5.45}$$

The next step comes from comparing equation (5.42) and equation (5.45). The former is fully equivalent to equation (5.41), which we know how to solve, and the latter is the one we are trying to solve. According to theorem 5.4, equations (5.42) and (5.45) will have the same roots if their coefficients are pairwise equal. Therefore, equation (5.44) will give us the roots of the canonical quadratic equation (5.45) if the following conditions apply:

$$p = a,$$
$$-2pg = b, \tag{5.46}$$
$$pg^2 - h = c.$$

We can determine the values of p, g, and h that satisfy these conditions. From the first one, we simply get $p = a$. We use this value in the second condition and obtain $g = -b/(2a)$. Finally, we use the values for p and g in the third condition and get $h = b^2/(4a) - c$. For these values of p, g, and h equations (5.41) and (5.45) are fully equivalent. Therefore, we can apply our solution (5.44) to equation (5.45), where we just need to express p, g, and h there through a, b, and c:

$$x = -\frac{b}{2a} \pm \sqrt{\frac{\frac{b^2}{4a} - c}{a}}. \tag{5.47}$$

It is customary to simplify this solution by using $2a$ as the common denominator. This produces the ubiquitous quadratic formula:

$$x = \frac{-b \pm \sqrt{b^2 - 4ac}}{2a}. \tag{5.48}$$

5.7 The Polynomial Remainder Theorem

This is a very important theorem for dealing with polynomials: it forms the basis for polynomial long division and is used in a proof that a polynomial of n-th degree has n roots. Its statement is as follows:

Theorem 5.6 *For any two polynomials f and g, such that $g \neq 0$, there exist unique polynomials q (the quotient) and r (the remainder), such that*

$$f = gq + r, \tag{5.49}$$

where either $r = 0$, or the degree of r is less than the degree of g.

Proof

This theorem comprises two statements: existence and uniqueness. We prove these two statements in turn.

1. *First, we prove the existence of polynomials q and r. We consider three cases:*
   (a) *Suppose $f = 0$. Then $q = r = 0$, and equation (5.49) is satisfied.*
   (b) *Suppose deg(f) <deg(g), where deg(f) is the degree of f. Then, we can select $q = 0, r = f$.*
   (c) *Suppose deg(f) ≥deg(g). Then, we prove the existence by induction on the degree of f.*
      i. *If deg(f) = 0, then deg(g) = 0, and both polynomials are equal to constants: $f = a; g = b$, where a, b are some nonzero numbers. Note that since $g \neq 0$ we conclude that $b \neq 0$. Then, we select $r = 0$ and $q = a/b$.*
      ii. *Next, we assume that q, r exist for deg(f) < n and consider the case deg(f) = n. Polynomials f and g have the form*

$$\begin{aligned} f &= a_n x^n + a_{n-1} x^{n-1} + ... + a_0, \\ g &= b_m x^m + b_{m-1} x^{m-1} + ... + b_0, \end{aligned} \tag{5.50}$$

where a_j, b_j are their coefficients and $m \leq n$. We form a new polynomial

$$h = f - a_n b_m^{-1} x^{n-m} g. \tag{5.51}$$

We substitute the explicit expressions for f and g into the definition of h:

$$\begin{aligned} h =& f - a_n b_m^{-1} x^{n-m} g \\ =& a_n x^n + a_{n-1} x^{n-1} + ... + a_0 - \\ & a_n b_m^{-1} x^{n-m} b_m x^m - a_n b_m^{-1} x^{n-m} b_{m-1} x^{m-1} - ... \\ & - a_n b_m^{-1} x^{n-m} b_0, \end{aligned} \tag{5.52}$$

where we observe that $a_n x^n = a_n b_m^{-1} x^{n-m} b_m x^m$ *and the highest order terms cancel. Therefore, polynomial* h *has a lower degree than polynomial* f*, and the theorem is true for this polynomial. Then there exist polynomials* q_h *and* r*, such that we can represent* h *as*

$$h = q_h g + r. \tag{5.53}$$

We substitute expression (5.51) for h *in the last equation:*

$$f - a_n b_m^{-1} x^{n-m} g = q_h g + r \tag{5.54}$$

and rearrange the terms:

$$f = \left(q_h + a_n b_m^{-1} x^{n-m}\right) g + r. \tag{5.55}$$

This is the desired expression $f = qg + r$ *if we denote*

$$q = q_h + a_n b_m^{-1} x^{n-m}. \tag{5.56}$$

This completes the induction.

2. *Uniqueness.*

We prove the uniqueness of q, r *by contradiction. Let's assume that there are two different pairs of polynomials* q *and* r*:*

$$\begin{aligned} f &\equiv g q_1 + r_1, \\ f &\equiv g q_2 + r_2. \end{aligned} \tag{5.57}$$

We also assume that $q_1 \neq q_2$ *or* $r_1 \neq r_2$*. We subtract these two equations to get:*

$$g(q_1 - q_2) \equiv (r_2 - r_1). \tag{5.58}$$

For equation (5.58) we consider two cases:

(a) *The degree of* g *is zero, and both remainders must be zero:* $r_1 = r_2 = 0$ *. Then, we have*

$$g(q_1 - q_2) \equiv 0. \tag{5.59}$$

Since $g \neq 0$*, equation (5.58) yields* $q_1 \equiv q_2$*, which (together with* $r_1 \equiv r_2$*) contradicts our assumption.*

(b) *The degree of* g *is nonzero. If* q_1 *is not identically equal to* q_2*, we compute the degrees of polynomials on both sides of equation (5.58). The degree on the left-hand side is greater or equal to the degree of* g*, but the right-hand side has the degree that is less than the degree of* g*. This can only happen if* $q_1 \equiv q_2$*. But then from equation (5.58) we get* $r_1 \equiv r_2$*. This contradicts our original assumption.*

Amazingly, operations on polynomials follow the same scheme as operations on real or complex numbers. Indeed, we can add, subtract, multiply and divide polynomials, and every time we will get another polynomial as a result, which is the property of closure. Moreover, addition and multiplication satisfy the properties of commutativity, associativity, and distributivity, just like numbers do. This is an example of a situation when seemingly different mathematical objects have the same traits, pointing to some fundamental truths that unite them.

5.8 The Fundamental Theorem of Algebra

We will get to the Fundamental Theorem of Algebra shortly. First, we need to prove another important theorem:

Theorem 5.7 *A polynomial of an n-th degree $P(x)$, where $n \geq 1$, has a root $x = a$ if and only if $P(x)$ can be represented as*

$$P(x) = (x - a)Q(x), \tag{5.60}$$

where the degree of polynomial $Q(x)$ is $n - 1$.

Proof

We prove the direct and converse statements in turn.[4]

1. *We prove that if $x = a$ is a root of $P(x)$, then $P(x)$ can be represented as in equation (5.60). We note that $(x - a)$ is a first-degree polynomial and therefore we can invoke the polynomial remainder theorem 5.6:*

$$P(x) = (x - a)Q(x) + r. \tag{5.61}$$

In this equation we set $x = a$ and recall that $P_n(a) = 0$. Then $r = 0$. We get

$$P(x) = (x - a)Q(x). \tag{5.62}$$

Since $P(x)$ has degree n and $x - a$ has degree 1, $Q(x)$ must have degree $n - 1$.

2. *We prove that if polynomial $P(x)$ can be represented as in equation (5.60), then $x = a$ is a root of $P(x)$. We substitute $x = a$ in equation (5.60) and this proves the statement.*

Now, we proceed to formulating the Fundamental Theorem of Algebra:

[4]See also exercise 19 for this chapter.

Theorem 5.8 *Every nonzero polynomial with degree $n \geq 1$ has at least one complex root.*[5]

In this section we accept this theorem without a proof. In section 7.16 we will discuss why this theorem is true. That argument, while convincing, will still fall a bit short of a complete and rigorous proof.

> The first rigorous proof of this theorem was by Jean-Robert Argand, an amateur mathematician. He is mentioned on p. 48 as the first person who displayed complex numbers on a plane.

From the Fundamental Theorem of Algebra follows another important theorem:

Theorem 5.9 *A nonzero polynomial of n-th degree has n complex roots (counted with multiplicity).*

Proof

Indeed, take a polynomial of n-th degree $P_n(x)$. According to the Fundamental Theorem of Algebra (theorem 5.8) it has at least one root. Then, according to theorem 5.7, it can be represented as

$$P_n(x) = (x - x_1)P_{n-1}(x), \tag{5.63}$$

where x_1 is a root, and $P_{n-1}(x)$ is a polynomial of $(n-1)$-th degree. Then $P_{n-1}(x)$ can be represented in a similar way, yielding

$$P_n(x) = (x - x_1)(x - x_2)P_{n-2}(x). \tag{5.64}$$

We continue this process until we get to a polynomial of zeroth degree, which is a constant.

Then any polynomial of n-th degree can be represented as

$$P_n(x) = c(x - x_1)(x - x_2)...(x - x_n), \tag{5.65}$$

where $x_1, x_2, ..., x_n$ are the roots of $P_n(x)$ and c is that polynomial of zeroth degree that we end up with after repeatedly applying theorems 5.8 and 5.7. This means that a cubic equation would have three roots, a quartic equation (that is, one that has a 4-degree polynomial equal to zero) – four roots and so on. There are formulas for solving cubic and quartic equations, although they

[5]Since a real number is a particular case of a complex number, a polynomial with one or more real roots also conforms to this theorem.

are very cumbersome. Interestingly, such formulas do not exist for a general polynomial of degree 5 and up (this has been proved).

The last statement does not mean that some particular higher-degree polynomial equations cannot be solved analytically. Some special cases do have analytical solutions. Consider the following polynomial equation:

$$x^8 - 1 = 0. \tag{5.66}$$

We can represent the left-hand side as a composite function:

$$\left(\left(x^2\right)^2\right)^2 = 1. \tag{5.67}$$

To solve for x, we successively compute square roots, each time keeping in mind that there are two possible signs in the right-hand side:

$$\begin{aligned} \left(x^2\right)^2 &= \pm 1 \\ x^2 &= \pm\sqrt{\pm 1} \\ x &= \pm\sqrt{\pm\sqrt{\pm 1}}. \end{aligned} \tag{5.68}$$

In the last equation, all $\pm$ signs are independent, producing the total of 8 combinations. Two of them, $\pm\sqrt{\sqrt{1}} = \pm 1$ produce real values $+1$ and -1, which are indeed roots of the original equation. The others produce complex values. This is an illustration of the fact that an eighth-power polynomial equation has eight roots.

Évariste Galois was a genius French mathematician who was killed in a duel at the age of 20. He discovered the conditions for a polynomial to be solved in radicals (that is, using the four arithmetic operations and rational exponents). The field of mathematics today called Galois theory shows, for example, that equation $x^5 - 1 = 0$ is solvable, but $x^5 - x - 1 = 0$ is not. In doing this research, Galois formulated what is now known as the group theory, which has become a large field of research in mathematics. His work was only appreciated posthumously.

Another genius mathematician who made important contributions to this field was Niels Henrik Abel, who died at the age of 26. Today, the most prestigious prize in mathematics bears his name.

Using the Fundamental Theorem of Algebra, we can tighten the condition for two polynomials to be equal. Above we proved that if two polynomials are identically equal ($P_1(x) \equiv P_2(x)$), then their coefficients are pairwise equal (see theorem 5.4). The statement of this theorem requires that $P_1(x) = P_2(x)$

for all values of x—that is, at an infinite number of points. Turns out, this is overkill, and we just need to require the two polynomials to be equal at $N+1$ points, where N is the larger degree of the two polynomials.

Theorem 5.10 *Let N be the larger of the degrees of two nonzero polynomials $P_1(x)$ and $P_2(x)$. Then $P_1(x)$ and $P_2(x)$ are identically equal and their coefficients are pairwise equal if the polynomials are equal at $N+1$ points.*

Proof

We consider two cases.

1. *Case $N = 0$. Then each polynomial is simply a constant, and the theorem states that for one value of x we have these constants equal. This proves the statement of the theorem for this case.*
2. *Case $N \geq 1$. The proof is by contradiction. We assume that $P_1(x) = P_2(x)$ at $N+1$ points, but they are not identically equal. Consider the difference of the two polynomials:*

$$Q(x) = P_1(x) - P_2(x). \tag{5.69}$$

From the treatment of polynomial addition in section 5.4, this is a polynomial of a degree not exceeding N, where N is the larger of the degrees of the two polynomials $P_1(x)$ and $P_2(x)$. According to our assumption ($P_1 \not\equiv P_2$), polynomial $Q(x)$ is nonzero. Therefore, according to theorem 5.9, it has N complex roots or fewer. But, we also see that it has zero values at $N+1$ values of x, which is a contradiction.

Theorem 5.11 *Two nonzero polynomials $P(x)$ and $Q(x)$ have the same roots if and only if there exists such nonzero number r that*

$$P(x) \equiv rQ(x). \tag{5.70}$$

Proof

We prove the direct and the converse statements in turn:

1. *We prove that if two polynomials P and Q have the same roots, then there exists such nonzero number r that equation (5.70) is true.*

   Consider polynomial $R = bP + cQ$, where b and c are nonzero constants. This polynomial will have N-th degree or lower—that is, $deg(R) \leq N$. For the sake of simplicity, we will assume that $deg(R) = N$, though the proof does not change substantially if $deg(R) < N$. If polynomials P and Q have the same roots, then polynomial R has the same N roots as P and Q for any values of b and c. We denote these roots $x_1, \ldots, x_N$. Let us select some value

x_a that is not a root of these polynomials and compute R at this value of x:

$$R(x_a) = bP(x_a) + cQ(x_a). \tag{5.71}$$

Next, we select b and c in such a way that the right-hand side of this equation becomes zero. We require

$$bP(x_a) + cQ(x_a) = 0, \tag{5.72}$$

which yields

$$b = -c\frac{Q(x_a)}{P(x_a)}. \tag{5.73}$$

Since x_a is not a root of $P(x)$, this value of b exists. Then for these particular values of b and c, polynomial R will have the $(N+1)$-th root at $x = x_a$. This may happen only if this polynomial is identically equal to zero. Therefore,

$$bP(x) + cQ(x) \equiv 0. \tag{5.74}$$

This yields

$$P(x) \equiv -\frac{c}{b}Q(x), \tag{5.75}$$

which is equivalent to equation (5.70) for $r = -c/b$.

2. *Now, we prove that if equation (5.70) is true for some nonzero r, then polynomials P and Q have the same roots. This directly follows from equation (5.70): if $Q(x)$ is zero for some x, then $P(x)$ is zero for these values of x and the other way around.*

From this theorem, we can prove another one:

Theorem 5.12 *Polynomials*

$$P(x) = a_n x^n + a_{n-1}x^{n-1} + \ldots a_0 \tag{5.76}$$

and

$$Q(x) = c(x - x_1)(x - x_2)\ldots(x - x_n) \tag{5.77}$$

are identical if and only if $c = a_n$ and values $x_1, \ldots, x_n$ are the roots of polynomial $P(x)$.

Proof

We need to prove two statements.

1. *First, we prove that if $c = a_n$ and values $x_1, \ldots, x_n$ are the roots of polynomial $P(x)$, then polynomials $P(x)$ and $Q(x)$ are identical.*

   Indeed, we can see that values $x_1, \ldots, x_n$ are roots of $Q(x)$ as well. According to theorem 5.11, there is such a number r that $P(x) \equiv rQ(x)$. If we expand all the parentheses in the definition of $Q(x)$ we shall see that the coefficient for the n-th power of x is c. Then, we must have $r = 1$ in $P(x) \equiv rQ(x)$, which completes the proof.

2. *Next, we prove that if polynomials $P(x)$ and $Q(x)$ are identical, then $c = a_n$ and values $x_1, \ldots, x_n$ are the roots of polynomial $P(x)$.*

   We get equation $c = a_n$ if we expand all the parentheses in the definition of $Q(x)$ and look at the coefficient for the n-th power of x. Separately, we recall theorem 5.11 to prove that these two polynomials have the same roots.

5.9 Vieta's Theorem

For simplicity we consider this theorem for quadratic polynomials.[6]

Theorem 5.13 *Coefficients of the quadratic equation*

$$ax^2 + bx + c = 0, \tag{5.78}$$

are related to the values of the roots:

$$\begin{aligned} x_1 + x_2 &= -\frac{b}{a}, \\ x_1 x_2 &= \frac{c}{a}. \end{aligned} \tag{5.79}$$

Proof

From theorem 5.12 we know that polynomials

$$ax^2 + bx + c \equiv a(x - x_1)(x - x_2), \tag{5.80}$$

are identical as long as x_1, x_2 are the roots of the quadratic equation

$$ax^2 + bx + c = 0. \tag{5.81}$$

If we distribute the parentheses in equation (5.80), we get

$$ax^2 + bx + c \equiv ax^2 - a(x_1 + x_2)x + ax_1x_2. \tag{5.82}$$

Since these polynomials are identical, their coefficients must be pairwise equal (theorem 5.4). This yields the statement of the theorem.

---

[6]Extension of Vieta's theorem to higher degrees is the subject of exercise 16.

Take a quadratic equation $ax^2 + bx + c = 0$ that has two real nonzero roots. Let's consider the effect of the signs of coefficients a, b, and c on the signs of the two roots of this equation. We are looking at three cases:

1. All the coefficients have the same sign (that is, $a, b, c > 0$ or $a, b, c < 0$). From $x_1 x_2 = c/a > 0$ in equations (5.79) we see that the roots must be either both positive or both negative. From $x_1 + x_2 = -b/a$ we conclude that the latter option is true—that is, $x_1 < 0, x_2 < 0$.
2. In the list of coefficients a, b, c, the sign changes once. That includes four options, such as $a, b > 0, c < 0$ or $a > 0, b, c < 0$. For each option, $x_1 x_2 = c/a < 0$, which means that there is only one positive root.
3. In the list of coefficients, the sign changes twice. That includes two options: $a > 0, b < 0, c > 0$ and $a < 0, b > 0, c < 0$. Then, we have $x_1 x_2 = c/a > 0$, and the roots must be either both positive or both negative. From $x_1 + x_2 = -b/a$ we conclude that the former option is true—that is, $x_1 > 0, x_2 > 0$.

We see a pattern: the number of positive roots seems to be linked to the number of sign changes in the list of coefficients. René Descartes proved a theorem that generalizes this pattern. Known today as the Descartes rule of signs, it states that the number of positive roots of a polynomial does not exceed the number of sign changes in the sequence of that polynomial's coefficients (where zero coefficients are omitted), and that the difference between these two numbers is always even.

Vieta's theorem becomes particularly elegant if $a = 1$. Then $x_1 + x_2 = -b$ and $x_1 \cdot x_2 = c$. That is, the sum of the roots is equal to the negative of the second coefficient of the quadratic equation, and the product of the roots is equal to the free term. In this form it is the basis of the factoring method to solve quadratic equations. For example, consider the equation

$$x^2 - 5x + 6 = 0. \tag{5.83}$$

We can guess that the roots of this equation are $x_1 = 2, x_2 = 3$ because this is the only pair of values whose sum is equal to 5 and whose product is equal to 6. While this method has a limited application in practice (combinations of coefficients that enable such lucky guesses are rare), it can serve as yet another basis for the proof of the quadratic formula:

Proof

We seek roots x_1, x_2 of the following quadratic equation:

$$x^2 + bx + c = 0. \tag{5.84}$$

According to theorem 2.19, there exist two numbers u and v such that

$$\begin{aligned} x_1 &= u + v, \\ x_2 &= u - v. \end{aligned} \tag{5.85}$$

From Vieta's theorem we already know that $x_1 + x_2 = -b$. Separately, by the last two equations we also have $x_1 + x_2 = 2u$. Therefore, $u = -b/2$. This is already halfway to finding the values of x_1 and x_2. If we find v, we will be done. To do that, we use the second equation of Vieta's theorem. From $x_1 x_2 = c$ we get

$$\begin{aligned} x_1 x_2 &= (u + v)(u - v) \\ &= u^2 - v^2 \\ &= \frac{b^2}{4} - v^2 \\ &= c. \end{aligned} \tag{5.86}$$

Then

$$v^2 = \frac{b^2}{4} - c \tag{5.87}$$

or

$$v = \pm\sqrt{\frac{b^2}{4} - c}. \tag{5.88}$$

The expressions for u and v yield:

$$\begin{aligned} x_{1,2} &= u \pm v \\ &= -\frac{b}{2} \pm \sqrt{\frac{b^2}{4} - c}. \end{aligned} \tag{5.89}$$

Using 2 as the common denominator produces a more familiar form of the quadratic formula in the case when the coefficient for x^2 is equal to one:

$$x_{1,2} = \frac{-b \pm \sqrt{b^2 - 4c}}{2}. \tag{5.90}$$

This formula can be extended to a more general case of a quadratic equation with the coefficient for x^2 equal to a, where a is nonzero, but not necessarily equal to 1. We just divide that equation by a and then apply formula (5.90). The result can be reduced to a more familiar form given by equation (5.35).

François Viéte, Seigneur de la Bigotiére (1540-1623), commonly known as Vieta, was a lawyer and an amateur mathematician. These two professions are not as different as it may seem: both require making precise and unambiguous statements.

Exercise 7 below shows that if $p + iq$ is a root of a polynomial with real coefficients, then its complex conjugate $p - iq$ is also a root. Therefore complex roots always come in complex conjugate pairs. This fact links the condition on the discriminant of a quadratic equation with the theorem about geometric and arithmetic means. Indeed, consider the equation

$$x^2 + bx + c = 0. \tag{5.91}$$

that has roots x_1 and x_2. We already know that $b = -(x_1 + x_2)$ and $c = x_1 x_2$. Next, we consider three cases:

1. Both roots are real and have the same signs. The discriminant must be nonnegative:

$$b^2 - 4c = (x_1 + x_2)^2 - 4x_1x_2 \geq 0. \tag{5.92}$$

We move $4x_1x_2$ to the right-hand side and extract the square root to get

$$|x_1 + x_2| \geq 2\sqrt{x_1x_2}. \tag{5.93}$$

Since x_1 and x_2 have the same signs, the absolute value of a sum is equal to the sum of absolute values:

$$\frac{|x_1| + |x_2|}{2} \geq \sqrt{|x_1||x_2|}. \tag{5.94}$$

We see that a nonnegative value of the discriminant is underpinned by the inequality for the arithmetic and geometric means (see theorem 3.17).

2. Both roots are real and have different signs. Then the discriminant $b^2 - 4c$ is positive because $b^2 \geq 0$ and $c = x_1x_2 < 0$.
3. Both roots are complex. We know that they must be complex conjugate: $x_1 = p + iq$; $x_2 = p - iq$, where $q \neq 0$. Then

$$\begin{aligned} b^2 &= (x_1 + x_2)^2 = (2p)^2 = 4p^2 \\ 4c &= 4x_1x_2 = 4(p^2 + q^2). \end{aligned} \tag{5.95}$$

We see that $b^2 < 4c$ and the discriminant is negative. We do expect this to be true, but it is still nice to see an explicit confirmation.

Exercises

1. (a) Prove the Brahmagupta–Fibonacci identity:

$$(a^2 + b^2)(c^2 + d^2) = (ac - bd)^2 + (ad + bc)^2. \tag{5.96}$$

(b) Use this identity to prove theorem 2.25.

2. Prove Candido's identity:

$$(a^2 + b^2 + (a+b)^2)^2 = 2(a^4 + b^4 + (a+b)^4). \tag{5.97}$$

3. Consider two polynomials $P_N(x)$ and $P_M(x)$ of degrees N and M. What is the degree of their product $P_N(x)P_M(x)$? Which theorem or theorems from this chapter can be used to determine the degree of $P_N(x)P_M(x)$?

4. Consider polynomial $P(x)$ with real coefficients and a complex value of x. Prove that $P(x^*) = (P(x))^*$, where the asterisk denotes a complex conjugate. (Hint: use induction to prove that $(x^n)^* = (x^*)^n$.)

5. In equation (5.25), the degree of the resulting polynomial is the same as the larger degree of the two polynomials in a sum. Give an example when the degree of a sum of two polynomials is lower than the degree of either of the two polynomials that are being added.

6. Prove that real roots of the following equation for x

$$(x - c^2 - c - 1)\left(x + \left(c^2 + 4c + \left|a + \frac{1}{a}\right|^2\right)\right) = 0, \tag{5.98}$$

have different signs for any values of parameters a and c.

7. Use exercise 4 to prove that complex roots of a polynomial with real coefficients always come in complex-conjugate pairs.

8. Exercise 7 implies that complex roots of a polynomial with real coefficients come in pairs of complex conjugates. Since a real number can be viewed as a particular case of a complex number (with the imaginary part equal to zero), the result of exercise 7 is valid for real roots of polynomials as well. Why then do the real roots not necessarily come in pairs?

9. Prove that if $a + b\sqrt{c}$ is a root of a polynomial with rational coefficients, where a, b are rational numbers and c is such an integer that $\sqrt{c}$ is irrational, then $a - b\sqrt{c}$ is also a root.

10. Consider two polynomial functions: $f(x)$ is a polynomial of degree n, and function $g(y)$ is a polynomial of degree m.

$$\begin{aligned} f(x) &= c_n x^n + c_{n-1} x^{n-1} + \ldots + c_0, \\ g(y) &= b_m y^m + b_{m-1} y^{m-1} + \ldots + b_0. \end{aligned} \tag{5.99}$$

(a) Use theorem 5.9 to determine the degree of polynomial $g(f(x))$.

(b) Is it true or false that $f(g(x)) = g(f(x))$?

(c) What is the condition for some of the roots of $g(f(x))$ to be the same as the roots of $f(x)$?

11. Use induction to prove that

$$(a-b)\sum_{n=0}^{N} a^n b^{N-n} = a^{N+1} - b^{N+1}. \tag{5.100}$$

12. Prove that for $k^2 < 4m$ equation $x^2 - 2nx + kn - m = 0$ has two distinct real roots for any value of n.
13. Consider a solution below. We seek x that satisfies the following equation:

$$x^3 + x^2 + x + 1 = 0. \tag{5.101}$$

It is obvious that $x = 0$ is not a root of this equation, so we can divide it by x. After that, we move one term to the right-hand side to get

$$x^2 + x + 1 = -\frac{1}{x}. \tag{5.102}$$

We substitute the left-hand side back into equation (5.101):

$$x^3 - \frac{1}{x} = 0, \tag{5.103}$$

which yields

$$x^4 = 1. \tag{5.104}$$

It is obvious that $x = 1$ is a solution of equation (5.104). However, this value does not satisfy the original equation (5.101).
Do the following:
    (a) Generalize this "solution" to polynomials for x in the form $\sum_{n=0}^{N} x^n$.
    (b) Go through a similar false way of solving equation $x + 1 = 0$.
    (c) How did the root $x = 1$ appear?
    (d) What are the correct solutions of equation (5.101)? (Hint: use the result of exercise 11.)
14. Prove that

$$\sum_{k=0}^{n} \binom{n}{k} (-1)^k = 0. \tag{5.105}$$

15. Consider

$$S = \sum_{k=0}^{n} \binom{n}{k} L^k, \tag{5.106}$$

where L is a natural number. What are the conditions for n and L in order for S to be prime?

16. We considered Vieta's theorem for the quadratic polynomial. Extend it to the case of a cubic polynomial $x^3 + a_2x^2 + a_1x + a_0 = 0$. Can you detect a pattern for polynomials of higher degrees?
17. When we considered equation (5.58) for case 2a on p. 119, we noted that both remainders r_1, r_2 must be identically equal to zero. Why is this true?
18. When considering case 2b for a nonzero degree of polynomial g on p. 119, we concluded that $q_1 \equiv q_2$. From which theorem we can obtain this conclusion?
19. The proof of theorem 5.7 states that polynomial $r = 0$ for $x = a$. However, by itself this is not sufficient to conclude that equation (5.62) is valid for all values of x. Why $r \equiv 0$ is true in this case?
20. Exercise 6 in chapter 3 states that for any a, b it is the case that $a^2 + ab + b^2 \geq 0$. Generalize this result: determine values of r, for which the expression $a^2 + rab + b^2$ is nonnegative for any a, b.
21. In section 5.4 we stated that addition and multiplication for polynomials have the properties of commutativity, associativity, and distributivity. Prove that this is the case for first-degree polynomials.
22. The proof of theorem 5.2 misses a case. Identify that case and complete the proof.
23. Formulate the necessary and sufficient conditions for the following identity to be true:

$$\left|\sum_{n=0}^{N} a_n x^n\right| \equiv \sum_{n=0}^{N} a_n |x|^n. \tag{5.107}$$

6

Power Law, Exponents, and Logarithms

The standard set of arrows in the quiver of applied mathematics includes a number of functions with well known properties that appear over and over again in various areas, from physics, to biology, to economics. We are already familiar with the properties of polynomials, which are widely used in many applications. In this chapter, we explore a group of functions that are as common as polynomials: the power law, exponents, and logarithms.

6.1 Integer Exponents

The definition of positive integer exponents is straightforward: x^2 is defined as $x \cdot x$, x^3 is defined as $x \cdot x \cdot x$, and so on. Below we will extend the values of exponents to all real numbers. We will do this step by step. We also want to do this organically, so as to be consistent with the rest of the stuff we already know. To do that, we define the *power law function.* We start from the positive integer exponents first:

$$f(x) = x^n, \tag{6.1}$$

where parameter n is a natural number. Properties of the power law function are as follows:

Theorem 6.1 *For any x and natural n, m it is the case that*

$$x^{n+m} = x^n \cdot x^m. \tag{6.2}$$

Proof

The left-hand and the right-hand sides of this equation contain the same number of factors and are equal due to the associativity of multiplication:

$$\begin{aligned} x^{n+m} &= (\underbrace{x \cdot x \cdot \ldots \cdot x}_{n+m}) \\ &= (\underbrace{x \cdot x \cdot \ldots \cdot x}_{n}) \cdot (\underbrace{x \cdot x \cdot \ldots \cdot x}_{m}) \\ &= x^n \cdot x^m. \end{aligned} \tag{6.3}$$

DOI: 10.1201/9781003456766-6

Theorem 6.2 *For any x, y and natural n it is the case that*

$$(xy)^n = x^n \cdot y^n. \tag{6.4}$$

A proof of this theorem is the subject of exercise 2 for this chapter.

Theorem 6.3 *For any x and natural n, m it is the case that*

$$(x^n)^m = x^{nm}. \tag{6.5}$$

Proof

The proof is by induction for m:

1. *For $m = 1$ we get:*

$$(x^n)^1 = x^{n \cdot 1}, \tag{6.6}$$

   which is true.

2. *We assume that the theorem is true for m and consider it for $m+1$. We use theorem 6.1 to get*

$$(x^n)^{m+1} = (x^n)^m x^n. \tag{6.7}$$

   Since the theorem is true for m,

$$(x^n)^m x^n = x^{nm} x^n. \tag{6.8}$$

   We again use theorem 6.1:

$$x^{nm} x^n = x^{mn+n}, \tag{6.9}$$

   which is equal to $x^{(m+1)n}$.

Theorem 6.4 *Function $f(x) = x^n$ is strictly monotonically increasing for nonnegative x and natural n.*

A proof of this theorem is the subject of exercise 21 for this chapter. Another proof uses the binomial expansion:

Proof

We consider two values $x_1 > x_2$. We write $x_1 = x_2 + D$, where D is positive. Consider the binomial expansion

$$x_1^n = (x_2 + D)^n = \sum_{k=0}^{n} \binom{n}{k} x_2^k D^{n-k}. \tag{6.10}$$

One term in this sum is x_2^n (the one that corresponds to $k = n$). All other terms are positive, so that they only increase the value of the sum. Therefore, $x_1^n > x_2^n$, which proves the theorem.

Just as we generalized some other operations from natural numbers to real, we can (and should) do so for exponentiation. Our first step is to add the zero and negative exponents. We already established that

$$x^{n+m} = x^n x^m. \tag{6.11}$$

Let's denote $k = n + m$. Then the last equation can be written as

$$x^k = x^n x^{k-n} \tag{6.12}$$

or

$$x^{k-n} = \frac{x^k}{x^n}, \tag{6.13}$$

where k, n are natural numbers and $k > n$ for now. This gives us a cue for how to define the power law function for nonpositive values of n: we simply extend the definition of exponentiation in a way that is consistent with equation (6.13). Now x^q is defined as the product of x with q factors if $q > 0$, and as a value that is consistent with equation (6.13) for $q \leq 0$. For example, if $k = 1$ and $n = 2$ in equation (6.13), we get

$$x^{-1} = \frac{x}{x^2} = \frac{1}{x}. \tag{6.14}$$

Similarly,

$$x^{-n} = \frac{1}{x^n}. \tag{6.15}$$

Finally,

$$x^0 = 1. \tag{6.16}$$

Note that for nonpositive integer exponents we must have $x \neq 0$.

> Newton's law of gravity states that the force between two bodies is proportional to $1/r^2$, and the energy of their interaction is proportional to $1/r$. Both are power law functions with a negative exponent for r.
>
> Newton was a Lucasian Chair of Mathematics in the University of Cambridge, and in recent times this position was held by the great Stephen Hawking. Professor Hawking famously expressed his opinion that these particular power law dependencies made it possible for the Universe to emerge in the Big Bang.

We have successfully extended the power law function to negative integer exponents. Can we now extend the exponent to noninteger values? To do that, we use the inverse functions for the power law.

6.2 Radicals as Inverse Exponents

We already know that a strictly monotonic function always has an inverse and that the power law function x^n for a nonnegative argument and natural exponent is strictly monotonic. Therefore, it has an inverse. Such an inverse is called a *root* or a *radical* and is denoted as $\sqrt[n]{x}$. We already considered the case of $n = 2$, when the inverse function is called the square root (see p. 36). By definition, for nonnegative values of x and natural n we have

$$\begin{aligned} \left(\sqrt[n]{x}\right)^n &= x, \\ \sqrt[n]{x^n} &= x. \end{aligned} \tag{6.17}$$

Since we have a convenient property given by theorem 6.3, we can introduce a new notation that is consistent with this property. We define a reciprocal power as

$$\sqrt[n]{x} = x^{\frac{1}{n}}. \tag{6.18}$$

Indeed, now equations (6.17) can be written as

$$\begin{aligned} \left(\sqrt[n]{x}\right)^n &= \left(x^{\frac{1}{n}}\right)^n = x^{\frac{n}{n}} = x, \\ \sqrt[n]{x^n} &= (x^n)^{\frac{1}{n}} = x^{\frac{n}{n}} = x. \end{aligned} \tag{6.19}$$

For radicals that are written as fractional powers, we can prove all the properties that are true for natural (nonfractional) exponents:

Theorem 6.5 *For any $a, b \geq 0$ and natural n it is the case that*

$$\sqrt[n]{ab} = \sqrt[n]{a} \cdot \sqrt[n]{b}. \tag{6.20}$$

Proof
This proof is by contradiction. Let

$$\sqrt[n]{ab} \neq \sqrt[n]{a} \cdot \sqrt[n]{b}. \tag{6.21}$$

We raise both sides of this inequality to power n. Since the power law function is strictly monotonic (see theorem 6.4), the inequality is preserved, and we get

$$\begin{aligned} ab &\neq \left(\sqrt[n]{a} \cdot \sqrt[n]{b}\right)^n \\ &= \left(\sqrt[n]{a}\right)^n \cdot \left(\sqrt[n]{b}\right)^n \\ &= ab, \end{aligned} \tag{6.22}$$

which is a contradiction.

Theorem 6.6 *For any $a \geq 0$ and natural n, m it is the case that*

$$\sqrt[n]{a^m} = \left(\sqrt[n]{a}\right)^m. \tag{6.23}$$

A proof of this theorem is the subject of exercise 5 for this chapter.

Theorem 6.7 *For any $a \geq 0$ and natural n, m it is the case that*

$$\sqrt[nm]{a} = \sqrt[n]{\sqrt[m]{a}} = \sqrt[m]{\sqrt[n]{a}}. \tag{6.24}$$

Proof

The proof is by contradiction. Let's assume that $\sqrt[nm]{a} \neq \sqrt[n]{\sqrt[m]{a}}$. We use the fact that the power law function is monotonically increasing for $a > 0$ (theorem 6.4) and raise this inequality to power mn:

$$\left(\sqrt[nm]{a}\right)^{mn} \neq \left(\sqrt[n]{\sqrt[m]{a}}\right)^{mn}. \tag{6.25}$$

In the left-hand side, we get a by definition. In the right-hand side, we use theorem 6.3 to get

$$\begin{aligned} a &\neq \left(\left(\sqrt[n]{\sqrt[m]{a}}\right)^n\right)^m \\ &= \left(\sqrt[m]{a}\right)^m \\ &= a. \end{aligned} \tag{6.26}$$

We obtain $a \neq a$; this contradiction proves the theorem.

Theorem 6.8 *For any $a > 0$ and natural n it is the case that*

$$\sqrt[n]{\frac{1}{a}} = \frac{1}{\sqrt[n]{a}}. \tag{6.27}$$

A proof of this theorem is the subject of exercise 3 for this chapter.

From the above proofs we see that properties of exponentiation apply to integer exponents or positive base values (or both). Failure to recognize this may lead to mistakes, as demonstrated by a flawed application of theorem 6.6:

$$\left((-1)^{\frac{1}{2}}\right)^2 = \left((-1)^2\right)^{\frac{1}{2}}. \tag{6.28}$$

The left-hand side of this equation computes to

$$\left((-1)^{\frac{1}{2}}\right)^2 = i^2 = -1. \tag{6.29}$$

However, the right-hand side computes to

$$\left((-1)^2\right)^{\frac{1}{2}} = (1)^{\frac{1}{2}} = 1. \tag{6.30}$$

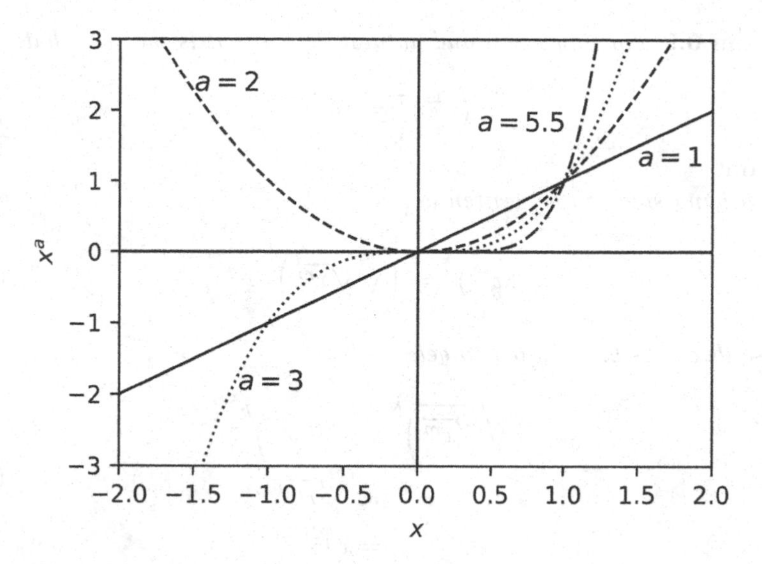

FIGURE 6.1
Plot of the power law function

This contradiction results from wrongly applying theorem 6.6 to a case of a negative base and a fractional exponent. For swapping the square root and the square we should keep in mind that while $(\sqrt{a})^2 = a$, we also have $\sqrt{a^2} = |a|$, which is *not* the same as a for $a < 0$. Similar difficulties arise for other fractional exponents and negative base values.

6.3 Rational Exponents

Using a combination of integer exponents and radicals, we can construct a value of a positive number raised to any power that is a rational number:[1] If we need to compute $x^{m/n}$, then we first compute x^m and then extract the n-th root. We can do it the other way around too: extract the root first and then raise to power m:

$$x^{\frac{m}{n}} = \sqrt[n]{x^m} = \left(\sqrt[n]{x}\right)^m . \tag{6.31}$$

Figure 6.1 shows plots of the power law function x^a for several values of a. Note that for integer values of a the domain of the function is all real numbers (including $x < 0$), but for $a = 5.5$ it is limited to $x \geq 0$.

All already familiar properties extend to rational exponents:

[1]As a reminder, a rational number is one that can be represented as $r = \frac{m}{n}$, where n, m are integers.

Theorem 6.9 *For any $x > 0$ and natural n, m, k, l it is the case that*

$$\left(x^{\frac{m}{n}}\right)^{\frac{k}{l}} = x^{\frac{mk}{nl}}. \tag{6.32}$$

Proof
The left-hand side can be written as

$$\left(x^{\frac{m}{n}}\right)^{\frac{k}{l}} = \left(\sqrt[l]{\sqrt[n]{x^m}}\right)^k. \tag{6.33}$$

We use theorems 6.6 and 6.7 to get:

$$\begin{aligned}\left(\sqrt[l]{\sqrt[n]{x^m}}\right)^k &= \left(\sqrt[ln]{x^m}\right)^k \\ &= \sqrt[ln]{x^{mk}} \\ &= x^{\frac{mk}{ln}}.\end{aligned} \tag{6.34}$$

Theorem 6.10 *For any $x > 0$ and natural n, m, k, l it is the case that*

$$x^{\frac{m}{n}+\frac{k}{l}} = x^{\frac{m}{n}} a^{\frac{k}{l}}. \tag{6.35}$$

Proof
Indeed, we have

$$\begin{aligned}x^{\frac{m}{n}+\frac{k}{l}} &= x^{\frac{ml+kn}{nl}} \\ &= \sqrt[nl]{x^{ml+kn}} \\ &= \sqrt[nl]{x^{ml} \cdot x^{kn}}.\end{aligned} \tag{6.36}$$

We use theorem 6.5 to obtain:

$$\begin{aligned}x^{\frac{m}{n}+\frac{k}{l}} &= \sqrt[nl]{x^{ml} \cdot x^{kn}} \\ &= \sqrt[nl]{x^{ml}} \cdot \sqrt[nl]{x^{kn}} \\ &= x^{\frac{ml}{nl}} x^{\frac{kn}{nl}} \\ &= x^{\frac{m}{n}} x^{\frac{k}{l}}.\end{aligned} \tag{6.37}$$

Finally, we extend the definitions of exponents to negative rational numbers. We define

$$x^{-\frac{m}{n}} = \frac{1}{x^{\frac{m}{n}}}. \tag{6.38}$$

For negative rational exponents we have the same properties valid. This means that all the properties of natural exponents are extended to rational exponents.

The power law function for positive rational exponents is monotonically increasing, just as it is increasing for natural powers:

Theorem 6.11 *Function* $f(x) = x^{\frac{m}{n}}$ *is strictly monotonically increasing for positive* $x, m,$ *and* n.

Proof

We use theorems 4.4 and 6.4. Functions x^m *and* $y^{\frac{1}{n}}$ *are monotonically increasing. Function* $f(x) = x^{\frac{m}{n}}$ *is a composite of these two functions, therefore it is also monotonically increasing by theorem 4.2.*

The power law function for a negative rational exponent is monotonically decreasing. The proof of this statement is the subject of exercise 24 for this chapter.

6.4 From Rational to Real Exponents

The final step is the extension of exponentiation to real values of the exponent. There are two ways to do that. One (more common) method relies on calculus in a substantial way and is beyond the scope of this book. The second one also uses some calculus concepts, but it can be explained on an intuitive level. The following is still a peek into calculus; do not be frustrated if it is not crystal clear.

We already know how to compute x^r, where r is a rational number. We want to define a way to compute that function for a value of r that is not rational. Any irrational number has a decimal representation. For example, $\pi = 3.14159265...$ We also realize that any truncated decimal representation (such as $3; 3.1; 3.14; 3.141$ or 3.1415) is a rational number.

Let's now compute 2^π. We start computing $2^3 = 8; 2^{3.1} \approx 8.5741877;$ $2^{3.14} \approx 8.8152409;$ $2^{3.141} \approx 8.8213533;$ $2^{3.1415} \approx 8.8244111$ and so on. We notice that consecutive values become ever closer to each other. They also become ever close to some number, which is *defined* as 2^π. In practice, we can compute that number to any desired accuracy by taking enough digits in the truncated representation of π in the exponent.[2] This completes our extension of the power law function to all real values of the exponent.

6.5 The Exponential Function

In the previous section, we considered the power law function:

$$f(x) = x^r. \tag{6.39}$$

[2]We return to real-valued exponents in a more rigorous way in section 9.7.

Notation $f(x)$ implies that this is a function of x, and r is a parameter.[3] The exponential function turns the above expression on its head: now the exponent is the function's argument, and the base is the parameter:

$$g(r) = x^r, \tag{6.40}$$

where we consider the dependence of g on r for a fixed value of x.

> The idea to consider the exponent, and not the base, as a variable for a function belongs to Leonhard Euler. He also introduced the current notation e for the constant that we discuss in the next section.

6.5.1 The Number e

There are a few fundamental constants in math, which appear over and over in various equations. One of them is π. Another one, just as common, is called the *base of natural logarithms* and is denoted as e. The approximate value of e is given by 2.718281828459045. It is an irrational number.

In the majority of practical applications, the exponential function uses the number e as its base: $f(x) = e^x$ (see figure 6.2). Using e as the base of the exponential function makes many equations simpler, not more complex. To see why the number e naturally appears in many problems, we need to explore its definition. Here is a qualitative, nonrigorous argument that defines this number and explores one of its applications.

Consider an investment that grows at a yearly rate r. An initial amount of P_0 will grow to $P_1 = P_0(1+r)$ in a year. In the second year, it is the value of P_1 that grows. We get

$$P_2 = P_1(1+r) = P_0(1+r)(1+r) = P_0(1+r)^2. \tag{6.41}$$

At the end of the third year, we get $P_3 = P_0(1+r)^3$ and so on:

$$P_n = P_0(1+r)^n. \tag{6.42}$$

Now let us compute the same assuming a continuous growth, which means that we look at the value of this investment at any point in time, not necessarily aligned with the end of a year. As the first step, let's shrink the step size from a year to a month, a week or a day. Let's assume that there are m smaller steps in a year. At each smaller step, the investment grows by a factor of $(1+\frac{r}{m})$. However, for every fixed investment term, there are proportionally more steps. We get at the end of the n-th year

$$P_n = P_0\left(1+\frac{r}{m}\right)^{nm}. \tag{6.43}$$

[3] We introduced the notion of parameters on p. 82.

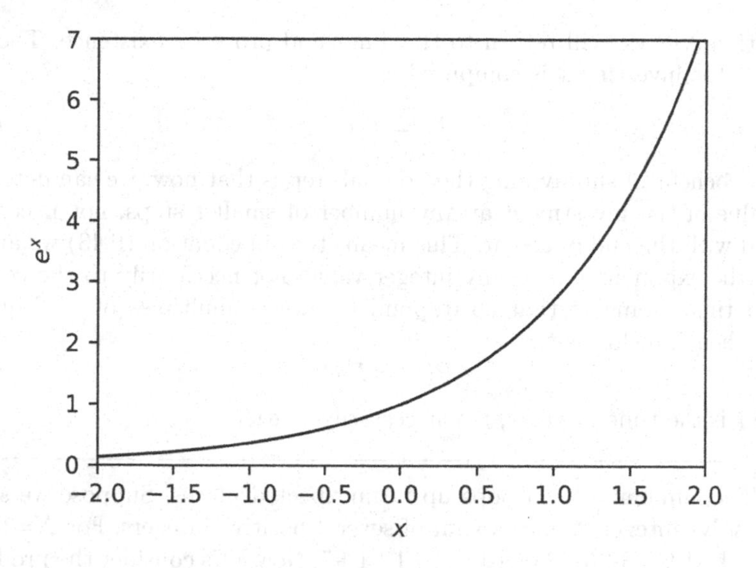

FIGURE 6.2
Plot of the exponential function

Here n is still the number of years, and m is the number of smaller steps in a year. Let us transform the last equation to isolate exponent nr:

$$P_n = P_0 \left[\left(1 + \frac{r}{m}\right)^{\frac{m}{r}} \right]^{nr}. \tag{6.44}$$

We can continue subdividing the steps, making each step an hour, or a second long. Then the value of m becomes very large (remember, it is the number of steps in a year). The value of $\frac{r}{m}$ is therefore small, and the value of $\frac{m}{r}$ is large. In the square brackets, we have a value of $(1+r/m)$ raised to the power m/r. Since the number of steps m is large, the base is close to 1, but it is raised to a large power. On one hand, if we raise a value that is close to 1 to some power, it will remain close to 1. On the other hand, if we raise a value that is greater than 1 to a very large power, it will be large. Therefore, in equation (6.44) we have two opposite, counteracting tendencies. Turns out, they balance each other, and for large m the result approaches neither 1 nor infinity:

$$\left(1 + \frac{r}{m}\right)^{\frac{m}{r}} \to 2.718281828459045... \tag{6.45}$$

The number e is usually defined as a limit of a sequence,[4] where the ratio m/r is combined in one variable:

$$\lim_{k\to\infty} \left(1 + \frac{1}{k}\right)^k = e \approx 2.718281828459045... \tag{6.46}$$

[4]We will formalize the notion of a limit in section 9.4.

In section 9.12 we will return to this limit and prove its existence. Then the value of the investment is computed as

$$P_n = P_0 e^{nr}. \tag{6.47}$$

The benefit of subdividing the original step is that now, we can determine the value of the investment at any number of smaller steps, not necessarily aligned with the end of a year. This means that in equation (6.43) we are free to set the exponent nm to any integer value, not necessarily to the coarsely spaced time moments that correspond to integer multiples of m. Equation (6.47) then transforms to:

$$P(t) = P_0 e^{rt}, \tag{6.48}$$

where t is the time in years, not necessarily integer.

> The number e often pops up in unexpected places. Suppose we split a positive integer N into a sum of several positive integers. For N=10 it could be $10 = 3+5+2$ or $10 = 1+1+1+7$. Now let's consider the product of these integers. For the above example, $3 \cdot 5 \cdot 2 = 30$ and $1 \cdot 1 \cdot 1 \cdot 7 = 7$. How do we select the integers to maximize that product? For N=10, the maximum is achieved for $3 \cdot 3 \cdot 2 \cdot 2 = 36$.
>
> Turns out, the general answer is to keep the value of all the terms as close to e as possible.

6.6 Properties of the Exponential Function

Now that we have defined the exponential function, we should investigate its properties. They stem from the already familiar properties for raising a number to a power. For example, we stated that $a^p a^q = a^{p+q}$. This and other properties of the power law function have counterparts for the exponential function:

$$\begin{aligned} e^{x+y} &= e^x e^y, \\ (e^x)^y &= e^{xy}, \\ e^{-x} &= \frac{1}{e^x}, \\ a^x b^x &= (ab)^x. \end{aligned} \tag{6.49}$$

> Consider a process that starts with some value of parameter x at time $t = 0$ and then we observe $x(t)$ as a function of time. Suppose that it has the following two properties:

1. It runs the same way whether it is starting now, has started in the past, or will start in the future. In other words, the run of this process does not depend on the time when we start the clock. Mathematically, this is formulated as follows: if $x(t)$ is a solution of a mathematical model, then $x(t - t_0)$, where t_0 is a time shift, is also a solution. All physical processes and many socioeconomic processes have this property. Basically, it states that laws of nature (or socioeconomic laws) do not change over time.
2. Suppose the process is *linear*. Mathematically, this means that if $x(t)$ is a solution, then $ax(t)$, where a is a constant, is also a solution. Many processes are linear or approximately linear. For example, propagation of electromagnetic waves is the same if we multiply the electromagnetic field strength by a constant factor, growth of a $1000 investment is proportionally 10 times larger than the growth of a $100 investment, and so on.

Processes that possess both these properties are surprisingly common. Importantly, the two properties above appear to be the two sides of the same coin. Namely, a time shift has the same effect on the solution as multiplying by a constant:

$$x(t - t_0) = ax(t). \tag{6.50}$$

The only function that satisfies this condition is the exponent. Indeed, if

$$x(t) = x_0 e^{\gamma t}, \tag{6.51}$$

then

$$\begin{aligned} x(t - t_0) &= x_0 e^{\gamma(t-t_0)} \\ &= e^{-\gamma t_0} x_0 e^{\gamma t} \\ &= a x_0 e^{\gamma t}, \end{aligned} \tag{6.52}$$

if we denote $a = e^{-\gamma t_0}$. The combination of these two properties is the primary reason for the wide spread of the exponential function in applications. Since trigonometric functions are close relatives of the exponent (see section 7.17), these properties also produce numerous trigonometric solutions.

In addition to electromagnetic waves or investment growth mentioned above, these types of solutions pop up in many applications. We see exponential solutions for radioactive decay, signal attenuation, the dynamics of chemical reactions, and in various problems in statistics. Trigonometric solutions describe countless oscillatory phenomena: pendulum motion, electromagnetic waves, acoustics, or the dynamics of satellite motion.

6.7 Is the Exponent Monotonic?

Above we proved that a power law function is strictly monotonic and therefore has an inverse. Here, we investigate whether the exponential function is also strictly monotonic.

Theorem 6.12 *The function e^x is strictly monotonically increasing in the domain of rational x.*

Proof

We consider two different values of the argument, $x_1 > x_2$. We need to prove that

$$e^{x_1} > e^{x_2}. \tag{6.53}$$

Since x_1 and x_2 are rational, we represent both as ratios of two integers:

$$e^{\frac{m}{n}} > e^{\frac{k}{l}}. \tag{6.54}$$

We multiply this inequality by the reciprocal of the right-hand side. Since the exponent and its reciprocal are positive, the sign of the inequality is preserved:

$$e^{\frac{m}{n}-\frac{k}{l}} > 1. \tag{6.55}$$

The difference of the rational numbers in the exponent is a rational number. Since $x_1 > x_2$, this number is positive. Let's denote it

$$\frac{m}{n} - \frac{k}{l} = \frac{p}{q}. \tag{6.56}$$

In the right-hand side of the inequality, we are free to replace the 1 *as follows:*

$$e^{\frac{p}{q}} > 1^{\frac{p}{q}}. \tag{6.57}$$

From theorem 6.11 and from $e > 1$ it follows that this inequality is true.

The exponential function is strictly monotonic in the domain of real numbers as well, but a proof of that is beyond the scope of this book. Intuitively, it is clear why the exponent should be a monotonically increasing function over all real numbers. An irrational number can be approximated to any desired accuracy by rational numbers. If the exponent is increasing over rational numbers, it is hard to expect that its value at some irrational value of x can "stick out" from the exponent values at all the rational numbers that tightly surround x.

Both the power law and the exponent are frequently used to model time-varying phenomena. Their behavior is different, but given enough time, the exponential growth always beats a power law. In fact, among the functions that are most commonly used in applications, the exponent is notable for its fast growth.

A striking example of that we encounter in cosmology. We already mentioned the Big Bang that has created our universe. However, this universe would have been quite small if not for a brief period of extremely fast exponential inflation very shortly after the Big Bang. According to some mathematical models, spacetime expanded by more than 10^{21} times over the period from 10^{-36} to 10^{-32} seconds!

6.8 Logarithms

Since the exponential function is strictly monotonic, it has an inverse. The inverse of the exponent is called the logarithm. When e is used as the base of the exponent, the inverse of the exponential function is denoted as $\ln x$ or $\ln(x)$. We omit the parentheses when it is clear what the argument of the logarithm is.

Below we investigate properties of the logarithm. Since the logarithm and the exponent are inverses of each other, we have:

$$\begin{aligned} e^{\ln x} &= x, \\ \ln e^{y} &= y. \end{aligned} \tag{6.58}$$

Values of e^x are always positive. The domain of the logarithm spans the range of e^x, therefore the argument of $\ln x$ must be positive.[5] Let's explore properties of this function. We start from computing $\ln(ab)$:

Theorem 6.13 *For any $a, b > 0$ it is the case that*

$$\ln(ab) = \ln a + \ln b. \tag{6.59}$$

Proof

We express a and b in a special way:

$$\begin{aligned} a &= e^{\ln a}, \\ b &= e^{\ln b} \end{aligned} \tag{6.60}$$

[5] To be exact, a logarithm of a negative number may be defined in complex numbers, somewhat similarly to the square root of a negative number, but this is beyond the scope of this book.

to get

$$\begin{aligned}\ln(ab) &= \ln\left(e^{\ln a}e^{\ln b}\right)\\ &= \ln\left(e^{\ln a+\ln b}\right).\end{aligned} \tag{6.61}$$

The "outer" logarithm and the exponent annihilate. We obtain:

$$\ln(ab) = \ln a + \ln b. \tag{6.62}$$

Theorem 6.14 *The logarithm of a ratio of two positive numbers is given by*

$$\ln\frac{a}{b} = \ln a - \ln b. \tag{6.63}$$

A proof of this theorem is the subject of exercise 11 for this chapter.

Theorem 6.15 *A logarithm of a^r for $a > 0$ is given by*

$$\ln a^r = r\ln a. \tag{6.64}$$

Proof

$$\begin{aligned}\ln a^r &= \ln\left(e^{\ln a}\right)^r\\ &= \ln e^{r\ln a}\\ &= r\ln a.\end{aligned} \tag{6.65}$$

We used the fact that the logarithm and the exponent are mutually inverse.

From the last property, we can see that the logarithm is a very slow growing function. (This is in contrast to the exponent, which grows very fast.) When we raise the argument to a power, the value of the logarithm is multiplied by that power:

$$\begin{aligned}&\ln 10 \approx 2.302585,\\ &\ln 10,000 \approx 2.302585\cdot 4 = 9.21034,\\ &\ln 100,000,000 \approx 2.302585\cdot 8 = 18.42068.\end{aligned} \tag{6.66}$$

Figure 6.3 shows a plot of $\ln x$.

To put a satellite on the orbit, we need to design a booster. The Tsiolkovsky rocket equation links the velocity of the exhaust V_e, the total velocity gained by the rocket V, the total starting mass of the rocket including propellant m_0 and the final mass of the rocket, when the propellant has been used m_f:

$$V = V_e \ln\frac{m_0}{m_f}. \tag{6.67}$$

In practice, the velocity of the satellite V must be four or five times greater that the velocity of exhaust V_e. This requirement translates into a very large value of the ratio m_0/m_f. The logarithm in the Tsiolkovsky rocket equation is the reason why space rockets must be so massive.

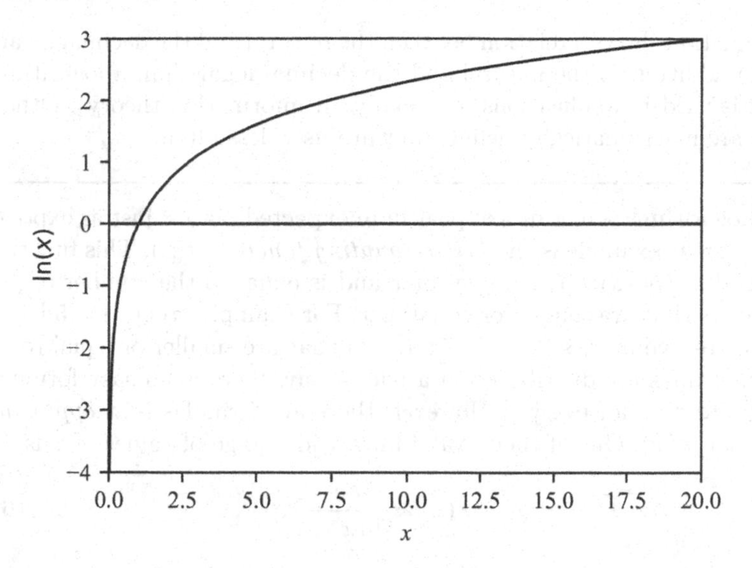

FIGURE 6.3
Plot of the logarithmic function

6.9 The Base of the Logarithmic Function

So far we have defined the logarithm as the inverse of the natural exponent $f(x) = e^x$. That logarithm is called the *natural logarithm*. There are other logarithmic functions that use other base values. For example, the inverse of $f_{10}(x) = 10^x$ is called the *decimal logarithm* or the *common logarithm*. (In spite of the name, use of the natural logarithm is more common, except in some areas of engineering dealing with the strength of electromagnetic or acoustic signals.) Since the decimal logarithm is the inverse of the 10-based exponent, we get

$$\log_{10} 10^x = x,$$
$$10^{\log_{10} y} = y. \tag{6.68}$$

There is a link between logarithms that use different base values. Let's apply the natural logarithm to both sides of the last equation:

$$\ln 10^{\log_{10} y} = \ln y. \tag{6.69}$$

In the left-hand side, we have a value raised to some power. We apply theorem 6.15 to move that power out as a multiplier:

$$\log_{10} y \cdot \ln 10 = \ln y. \tag{6.70}$$

This equation shows a relation between the natural and the decimal logarithms of y. In addition to the natural and the decimal logarithm, a logarithm with base 2 is used in applications (especially, in information theory). Other base values are mathematically defined but are used less often.

Logarithms sometimes appear in unexpected places, just as exponents do. A great example is the *prime-counting function* $\pi(x)$. This function is defined for positive real values of x and is equal to the number of prime numbers that are smaller or equal to x. For example, $\pi(11.3) = 5$ because there are five primes (2, 3, 5, 7 and 11) that are smaller or equal to 11.3. Since primes are distributed in a tricky way, there is no easy formula to compute $\pi(x)$ for every x. However, there are formulas that *approximate* function $\pi(x)$. One of them, valid for a wide range of values of x is

$$\pi(x) \approx \frac{x}{\ln x - 1}. \tag{6.71}$$

For example, plugging x = 1,000,000 in this formula produces $\pi(x)$ ≈78,030, compared with the exact value 78,498. This is a good approximation for a value that is difficult to obtain using other means. Somehow, a logarithm popped up in the approximation of a seemingly unrelated prime counting function!

We see that $\pi(x)$ grows slower than x, but since logarithm is a very slowly growing function, we still have a lot of primes for the values of x that are not crazy large. As a result, it is easy to form a product of several prime factors, but there is no easy and quick way to factor a composite number if that number is large because sifting through all candidate primes becomes too computationally expensive. This provides a foundation for modern encryption algorithms: to decode a message, we need to know the factors in the number that was used for encryption.

Approximations for the prime counting function were first guessed by Adrien-Marie Legendre and Peter Gustav Lejeune Dirichlet. The breakthrough in the derivation of formula (6.71) came in the second half of the 19th century due to use of the Riemann zeta function in the complex plane. (We will talk about this function in exercise 23 for chapter 9.) In the 20th century, Atle Selberg and Paul Erdős independently came up with "elementary proofs" for formula (6.71). While not elementary at all in the customary sense of this word, these proofs do not use advanced concepts like the zeta function in the complex plane.

Exercises

1. A proof on p. 42 states that if $|a| < 1$, then consecutive powers $|a|^n$ can be made arbitrarily small. This means that if we take any positive value δ, there exists such a value of n that for all $m \geq n$ it is the case that $|a|^m < \delta$. Prove that this is indeed true.
2. Prove theorem 6.2.
3. Prove theorem 6.8.
4. Theorem 6.3 states that $(x^n)^m = x^{nm}$. It is proved above by induction for m. Provide a proof by induction for n.
5. Prove theorem 6.6 by induction for m.
6. From $\ln p^r = r \ln p$ prove that $\ln 1 = 0$. Use two methods: (a) consider a special value for p, and (b) consider a special value for r.
7. From $\ln ab = \ln a + \ln b$ prove that $\ln 1 = 0$.
8. Prove that for positive a, b and rational r it is the case that
$$\left(\frac{a}{b}\right)^r = \frac{a^r}{b^r}. \tag{6.72}$$
9. Prove that $\log_{10} 2$ is irrational.
10. Prove that the natural logarithm of a rational number is irrational. (Hint: use a proof by contradiction and use the fact that number e is irrational.)
11. Prove theorem 6.14.
12. The proof of theorem 6.13 above manipulates the expression $\ln(ab)$ to arrive at $\ln a + \ln b$. Note that for $a < 0, b < 0$ the starting expression $\ln(ab)$ is defined, but the final expression $\ln a + \ln b$ is not. Where in the proof we have used the condition $a > 0, b > 0$?
13. Prove that for positive numbers a, b the condition $\ln(a+b) = \ln a + \ln b$ is true if and only if $(a-1)(b-1) = 1$.
14. Prove that if $\ln(ab) = \ln a \cdot \ln b$, then either both a and b are greater than e, or both are smaller than e. Can you detect a commonality with exercise 13?
15. Prove that for $a, b > 0$ it is the case that $a^{\ln b} = b^{\ln a}$.
16. Prove that for any positive numbers a, b it is the case that
$$\ln a^{\ln a} - \ln b^{\ln b} = \ln ab \cdot \ln \frac{a}{b}. \tag{6.73}$$
17. Are equations $\ln a^2 = \ln b^2$ and $2 \ln a = 2 \ln b$ equivalent? Why?

18. The associative property for addition applies to exponents of a number and is consistent with multiplying exponentiated numbers. Indeed:

$$\begin{aligned} a^{p+(q+r)} &= a^p a^{q+r} \\ &= a^p(a^q a^r) \\ &= (a^p a^q)a^r \\ &= a^{p+q}a^r \\ &= a^{(p+q)+r}. \end{aligned} \tag{6.74}$$

Prove that the distributive, commutative, and multiplication associative properties for the exponents are similarly consistent with the rules for operating with exponentiated numbers.

19. A student is solving the following equation:

$$\ln(ax^2) = \ln(bx + c). \tag{6.75}$$

She exponentiates this equation and collects the terms to get

$$ax^2 - bx - c = 0. \tag{6.76}$$

Then she uses the quadratic formula to obtain

$$x_{1,2} = \frac{b \pm \sqrt{b^2 + 4ac}}{2a}. \tag{6.77}$$

In the next exercise, she has to solve a slightly different equation:

$$\ln(ax^2 - bx) = \ln c. \tag{6.78}$$

She follows the same logic to arrive at the same answer. Are these two equations and their solutions equivalent?

20. Assume that $\ln x$ is a concave function—that is, for $x_1 \neq x_2$, $x_1, x_2 > 0$, and $0 < a < 1$ it is the case that

$$\ln(ax_1 + (1-a)x_2) > a\ln x_1 + (1-a)\ln x_2. \tag{6.79}$$

Use inequality (6.79) to prove that the arithmetic mean of x_1, x_2 is greater than the geometric mean of these two numbers.

21. Use induction to prove theorem 6.4.

22. Prove that if $0 < r < 1$, then $f(x) = r^x$ is monotonically decreasing.

23. In the proof of theorem 6.7, we mentioned that the power law function is monotonic. Why was that necessary?

24. Prove that the power law function x^r for the positive value of x and negative rational value of r is monotonically decreasing.

25. A student wants to use theorem 3.2 to prove that if $a \geq b$ and $c \geq d$, then $ac \geq bd$. She represents variables a, b, c, and d as:

$$\begin{aligned} a &= e^{\ln a}, \\ b &= e^{\ln b}, \\ c &= e^{\ln c}, \\ d &= e^{\ln d}. \end{aligned} \tag{6.80}$$

Since both the logarithm and exponent are monotonic increasing functions, she gets:

$$ac = e^{\ln a + \ln c} \geq e^{\ln b + \ln d} = bd. \tag{6.81}$$

While this proof achieves the desired result $ac \geq bd$, it imposes certain constraints on the values of a, b, c, and d. What are these constraints and can they be removed?

26. Find a flaw in the following proof. We are going to prove by induction that for any $a > 0$ and for any nonnegative integer n it is the case that $a^n = 1$. The two steps of induction are:

    (a) For $n = 0$ we do have $a^n = 1$.

    (b) For $n + 1$ we write:

$$a^{n+1} = \frac{a^n a^n}{a^{n-1}} = \frac{1 \cdot 1}{1} = 1. \tag{6.82}$$

27. Prove theorem 2.20 from theorem 2.19 using logarithms and exponents.

7

Trigonometry

7.1 How to Use Algebra for Solving Problems in Geometry

Geometry studies properties and relations of points, lines, and shapes. This is probably the oldest branch of math. René Descartes was among the first to realize that problems in geometry can be solved using the powerful apparatus of algebra. He observed that we can define the position of a point on a plane or in the 3D space by specifying its coordinates. For example, a point on a plane can be defined by a pair of coordinates x, y. These coordinates are just regular numbers. Then any relationship between two points x_1, y_1 and x_2, y_2 becomes an *algebraic* relationship between the two pairs of numbers x_1, y_1 and x_2, y_2. Now a geometric problem can be formulated in algebraic terms. Because of the power of algebra, many geometry problems are solved today by transitioning them into the domain of algebra.

> Descartes left his native France and for many years lived in the Dutch Republic. One of the reasons for this was that in the Netherlands he did not have as many distractions as he did in Paris, and he could concentrate on his studies. Not knowing the Dutch language helped.

In this connection between algebra and geometry, a great role is played by *trigonometry*. This is a topic in mathematics that algebraically connects side lengths and angles of triangles. To be sure, trigonometry has been known and used long before Descartes. But when we try to solve a geometric problem using algebraic methods, trigonometry always pops up.

7.2 Measuring Angles

By convention, angles are measured from the horizontal axis counterclockwise. Such angles are positive. If an angle is measured clockwise, its measure is

DOI: 10.1201/9781003456766-7

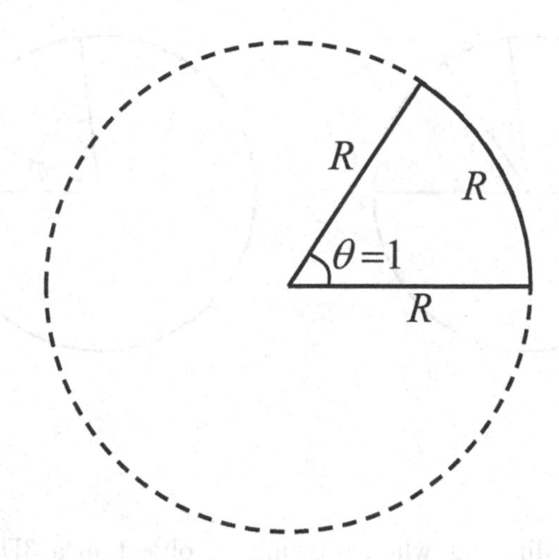

FIGURE 7.1
Definition of the radian

negative. In mathematics angles are measured in *radians*. The angle measure in radians is the ratio of the arc length to the radius of the circle. One radian corresponds to the angle, for which the length of the arc is equal to the radius of the circle (figure 7.1).

The circumference of a circle is computed as $2\pi r$, where r is the radius, so for the full circle the ratio of the arc to the radius is equal to 2π. Therefore, the full circle (or 360°) is 2π radians. A half-circle (or 180°) is π radians, and 90° is $\pi/2$ radians. Use of degrees or other units is mostly relegated to presenting the final result in a more digestible way.

Angles are not limited to a fraction of a circle or even to a full circle. They can have any real value. A large angle just means that an object is rotated by a large angle. For example, the Earth rotates by 360° or 2π radians approximately every 24 hours. In 2 days, the Earth will rotate by approximately 4π radians, and in a month by about 60π radians.

7.3 Adding and Subtracting Angles

Angles are added or subtracted according to regular algebraic rules. On the unit circle, this corresponds to applying consecutive rotations clockwise or counterclockwise.

For adding angles on a plane we use the regular commutative property $\theta_1 + \theta_2 = \theta_2 + \theta_1$. Let's rotate a radius of a circle first by angle θ_1 and then by θ_2 and compare it with rotating first by θ_2 and then by θ_1. We will get the same result (figure 7.2).

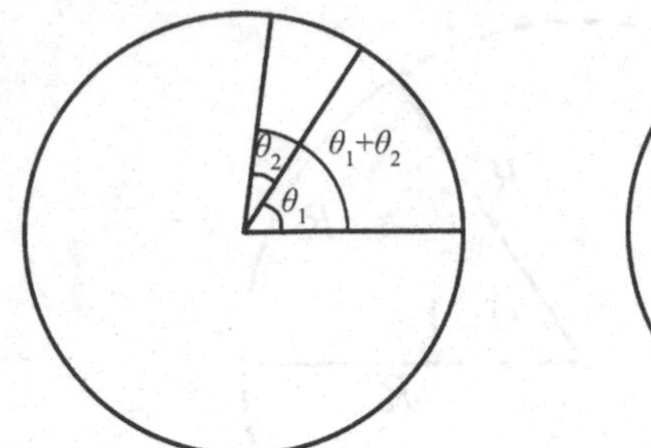

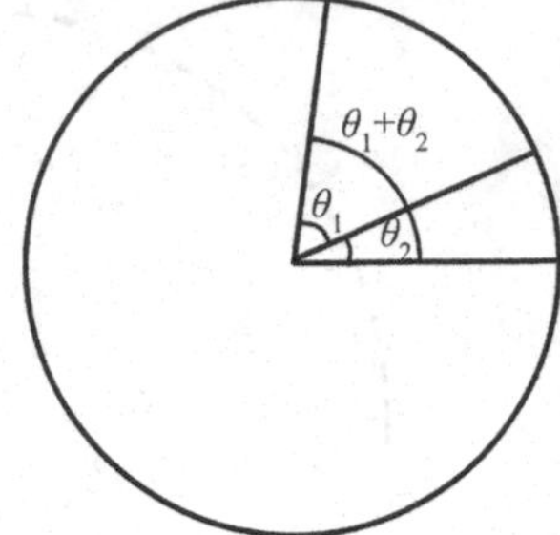

FIGURE 7.2
Adding angles

The situation is different when rotating an object in a 3D space. If we rotate an object around the same axis, the result does not depend on the order of rotations. However, this is not true for two rotations with respect to two different axes! Rotations in 3D are not commutative.[1] We will consider consequences of this fact on p. 171.

7.4 The Sine and Cosine Functions

Angles and sides of triangles are linked using the so-called *trigonometric functions*. There are six trigonometric functions. We start from two most common ones, the *sine* and the *cosine*. We consider horizontal and vertical coordinate axes (such coordinates are called Cartesian in honor of Descartes) and draw a radius of a circle, as shown in figure 7.3. The sine is a function that is defined as follows:

1. The range is all real numbers.
2. The domain are all numbers from -1 to 1 (inclusive).
3. The value of the input (that is, the angle θ) is measured in radians counterclockwise from the horizontal axis. The value of the sine is the vertical coordinate of the end point of the radius that forms angle θ with the horizontal axis. In terms of the right triangle on the figure, the sine function is the ratio of the opposite leg of the triangle to the hypotenuse.

[1]To convince yourself that this is true, take an asymmetric object and try to rotate it with respect to two different axes. Then do the same, but switch the order of these two rotations. The final positions of the object in these two experiments will be different.

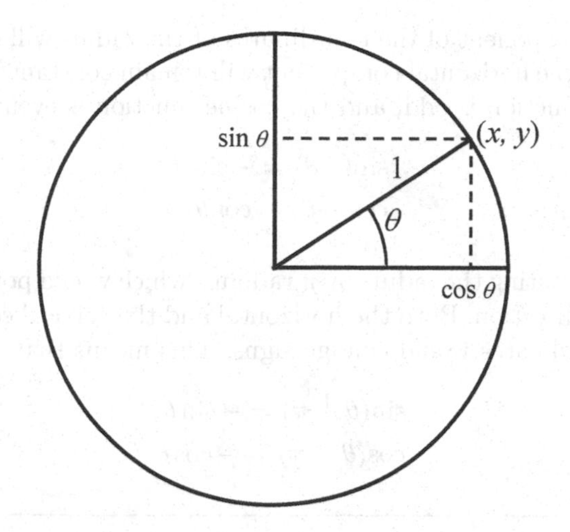

FIGURE 7.3
Sine and cosine

The notation for the sine function is $\sin\theta$ (see figure 7.3). In practice, there are better ways to compute the sine function than to draw a unit circle and measure the leg of a triangle.

This function is not algebraic. This means that it cannot be computed by applying a finite number of such operations as $+, -, \times, /$, raising a variable to an integer power, and radicals.

Cosine is a similar function: its value is the horizontal coordinate of the end point of that radius in figure 7.3. In terms of the right triangle on the figure, the cosine function is the ratio of the adjacent leg of the triangle to the hypotenuse. The notation for the cosine function is $\cos\theta$ (see figure 7.3). It is not algebraic either.

From the definition of sine and cosine, we observe the following properties of these functions:

1. Consider two angles that differ by an integer number of full rotations—that is, by $2\pi n$. In this case the radius ends up at the same place for both angles. Then values of the trigonometric functions must be the same for these two angles. Therefore,

$$\begin{aligned} \sin\theta &= \sin(\theta + 2\pi n), \\ \cos\theta &= \cos(\theta + 2\pi n). \end{aligned} \tag{7.1}$$

   We say that sine and cosine are periodic, and their period is equal to 2π.

2. Imagine rotating the radius in figure 7.3 not counterclockwise, but clockwise, which would correspond to a change in the sign of θ. The

vertical component of the coordinates of the radius will change the sign, and the horizontal component will remain constant. Therefore, the sine function is odd, and the cosine function is even:

$$\begin{aligned} \sin(-\theta) &= -\sin\theta, \\ \cos(-\theta) &= \cos\theta. \end{aligned} \tag{7.2}$$

3. Imagine rotating the radius by π radians, which would point it in the opposite direction. Both the horizontal and the vertical components of its coordinates would change signs. This means that

$$\begin{aligned} \sin(\theta+\pi) &= -\sin\theta, \\ \cos(\theta+\pi) &= -\cos\theta. \end{aligned} \tag{7.3}$$

Applications of trigonometry are not limited to solving problems in geometry. In contrast to many other functions, sine and cosine functions are oscillatory and do not grow indefinitely as their argument becomes large. When sine and cosine are used as a function of time, they become a good choice for describing such systems, where time is growing indefinitely, but some quantity as a function of time remains bounded. This, for example, includes electronic circuits, which operate continuously, but where the electrical current is limited by the technical specifications of the equipment. It is common to model electrical current or a telecommunication signal as $E\sin(2\pi f t)$, where t is time, E is called the amplitude, and f is the frequency of oscillations.

7.5 Most Common Trigonometric Identities

In this section we introduce the two most common trigonometric identities. Let's look at the right triangle in figure 7.3.

1. The Pythagoras theorem states that in a right triangle the sum of squares of the two legs is equal to the square of the hypotenuse. This leads to the so-called Pythagoras identity for trigonometric functions:

$$\sin^2\theta + \cos^2\theta = 1. \tag{7.4}$$

Consider angles beyond the first quadrant. The sine and cosine will still correspond to the legs of a right triangle, though the value of each function is then not necessarily positive. Since the Pythagoras identity uses squares of a sine and cosine, it is not affected by the signs of these functions. We see that the Pythagoras identity is valid for any values of θ.

2. The sum of nonright angles of a right triangle is equal to $\pi/2$. Therefore, if one angle is θ, the other one is $\pi/2 - \theta$. For angle θ and a triangle with a unit hypotenuse, the value of $\sin\theta$ is the length of the opposite leg, and $\cos\theta$ is the length of the adjacent leg. For the other nonright angle $\pi/2 - \theta$, the adjacent and the opposite legs swap their roles. Therefore, the length of the opposite leg for the angle θ is also the length of the adjacent leg for the angle $\pi/2 - \theta$. This leads to the *complementary angle identities*:

$$\begin{aligned} \sin\left(\frac{\pi}{2} - \theta\right) &= \cos\theta, \\ \cos\left(\frac{\pi}{2} - \theta\right) &= \sin\theta. \end{aligned} \tag{7.5}$$

This identity is clear from figure 7.3 if the argument of sine and cosine is between zero and $\pi/2$, where both functions are positive. For θ values outside of this interval, this identity is less obvious because we have to deal with different combinations of the signs of sine and cosine (and, unlike the case of the Pythagoras identity, the signs matter now). Instead of considering separately values of θ in all four quadrants, we will prove this identity for arbitrary values of θ in a different way in section 7.10.

7.6 Inverse Trigonometric Functions

The plot of the sine function is shown in figure 7.4. We already know that sine and cosine are periodic functions:

$$\begin{aligned} \sin(2\pi n + \theta) &= \sin\theta, \\ \cos(2\pi n + \theta) &= \cos\theta. \end{aligned} \tag{7.6}$$

For any value of θ, we get only one value of sine or cosine. However, for any value of sine and cosine between -1 and 1 we get infinitely many possible values of θ. The trigonometric functions do not pass the "horizontal line test." This means that they do not have inverses.

To avoid the problem with the horizontal line test, the inverses are not defined for sine or cosine proper, but for certain segments of these functions, where the function is monotonic and therefore where it passes the horizontal line test. Specifically, the inverse for $\sin\theta$ is defined for $-\pi/2 \leq \theta \leq \pi/2$, and the inverse for $\cos\theta$ is defined for $0 \leq \theta \leq \pi$.

Applying the inverse to the equation $\sin\theta = a$ where $-1 \leq a \leq 1$, does produce one solution $\theta = \sin^{-1} a$, which always lies in the interval $-\pi/2 \leq$

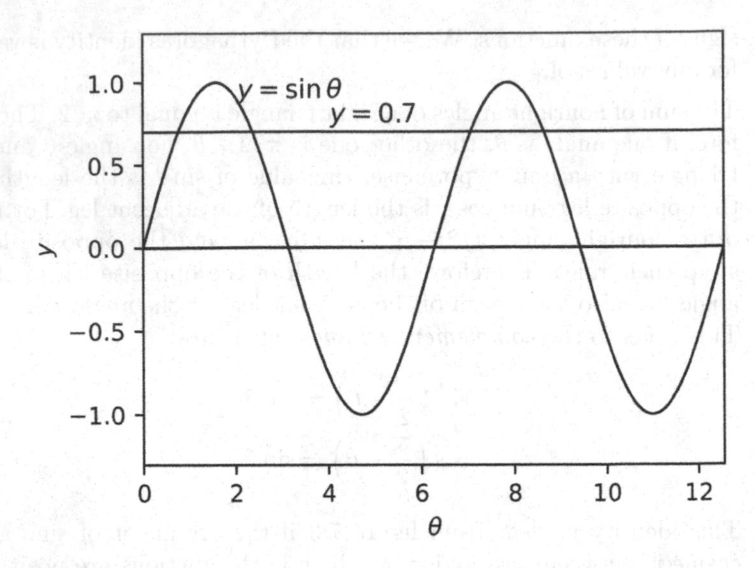

FIGURE 7.4
The sine function and a constant

$\theta \le \pi/2$. However, it misses all other solutions that can be seen in figure 7.4 as the intersections of the horizontal line and the plot of the sine. The following formula yields all solutions in one compact form. If $\sin\theta = a$, then

$$\theta = (-1)^n \sin^{-1} a + \pi n. \tag{7.7}$$

Here n enumerates the solutions and can have any integer value. If $n = 0$, we get the "main" solution $\theta = \sin^{-1} a$. For $n \neq 0$ we get other solutions that correspond to different intersections of the horizontal line and the plot of the sine. (Just be aware that for $a = \pm 1$ equation (7.7) lists each solution of $\sin\theta = a$ twice.)

For the cosine, the situation is similar: it also does not pass the horizontal line test. The inverse of $\cos\beta$ is defined for $0 \le \beta \le \pi$. If $\cos\beta = b$, then

$$\beta = \pm\cos^{-1} b + 2\pi n. \tag{7.8}$$

All solutions of the equation $\cos\beta = b$ are given by the combination of a sign (plus or minus in the right-hand side) and a value of n. (Similarly, for $b = \pm 1$ equation (7.8) lists each solution of $\cos\theta = b$ twice.)

7.7 Other Trigonometric Functions

A right triangle has three sides, from which one can form six ordered pairs. The ratios of the lengths of the sides in these pairs correspond to the six

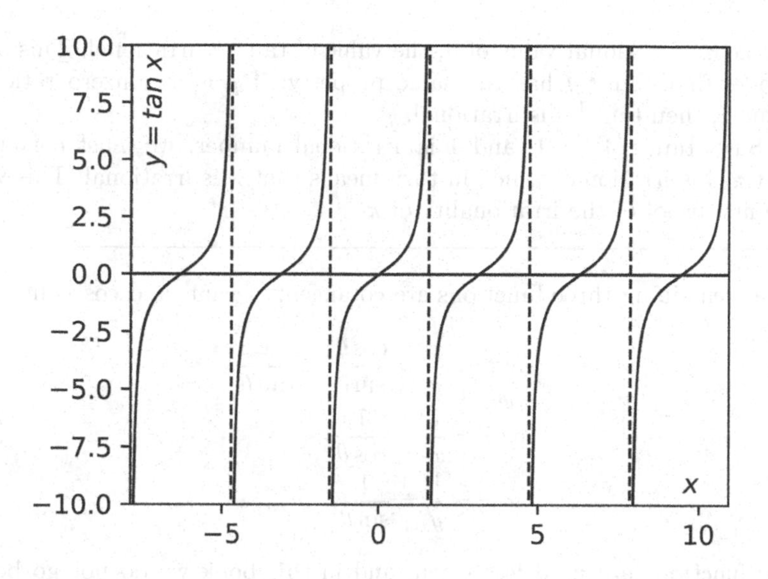

FIGURE 7.5
Plot of the tangent function

trigonometric functions. We already defined two of them—the sine and the cosine. Another one is the *tangent*. On a unit circle, tangent is the ratio of the vertical coordinate to the horizontal one (see figure 7.3). This means that

$$\tan\theta = \frac{y}{x} = \frac{\sin\theta}{\cos\theta}. \tag{7.9}$$

We know that both the sine and cosine just change their signs if we increment the argument by π (equations (7.3)). Then from equation (7.9) we see that $\tan\theta$ will retain its value (including the sign) if we increment its argument by π:

$$\tan(\theta+\pi) = \frac{\sin(\theta+\pi)}{\cos(\theta+\pi)} = \frac{-\sin(\theta)}{-\cos(\theta)} = \tan\theta. \tag{7.10}$$

We conclude that tangent is periodic with the period π. (As a reminder, sine and cosine are periodic with the period 2π.) Figure 7.5 shows a plot of the tangent function. Tangent is not defined at the points where $\cos\theta = 0$—that is, for $\theta = \pi/2 + n\pi$. The range of the tangent function is all real values, therefore, the domain for the inverse tangent is all real values. The range of the inverse tangent is the open interval from $-\pi/2$ to $\pi/2$.

Trigonometric functions indiscriminately apply to all real numbers in their respective domains, and it is hard to expect seeing any particular properties for computing a trigonometric function of, say, a rational number. Yet, in 1761 Johann Heinrich Lambert was able to prove that for any nonzero rational value of θ, the value of $\tan\theta$ is irrational. This also implies that $\tan^{-1} y$ has the same property: if y is a nonzero rational number, then $\tan^{-1} y$ is irrational.

Since $\tan(\pi/4) = 1$, and 1 is a rational number, we must conclude that $\pi/4$ is irrational, which in turn means that π is irrational. This was the first proof of the irrationality of π.

The remaining three functions are cotangent, secant, and cosecant:

$$\begin{aligned} \cot\theta &= \frac{x}{y} = \frac{\cos\theta}{\sin\theta} = \frac{1}{\tan\theta}, \\ \sec\theta &= \frac{1}{x} = \frac{1}{\cos\theta}, \\ \csc\theta &= \frac{1}{y} = \frac{1}{\sin\theta}. \end{aligned} \tag{7.11}$$

These functions are used less often, and in this book we do not go beyond providing their definitions.

7.8 Polar Coordinates

Cartesian coordinates is not the only way to specify a point on a plane. Another useful tool are *polar coordinates.* For any point on the plane (except the origin), we draw a line from the origin to that point. Then the position of that point is defined as the distance to the origin R and the angle θ that is formed by the ray with the x axis (figure 7.6.)

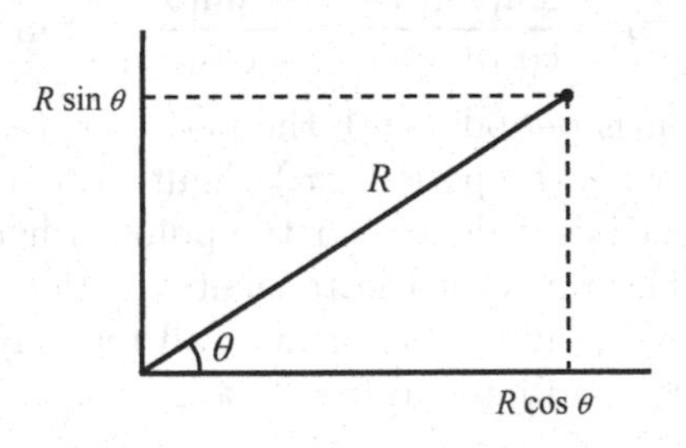

FIGURE 7.6
Polar coordinates

From the very definitions of trigonometric functions we obtain a transformation from polar to Cartesian coordinates:

$$\begin{aligned} x &= R\cos\theta, \\ y &= R\sin\theta. \end{aligned} \tag{7.12}$$

The inverse transformation is a bit more complicated. It is straightforward to compute the distance as

$$R = \sqrt{x^2 + y^2}. \tag{7.13}$$

For the angle, the first impulse is to compute it as $\theta = \tan^{-1} y/x$. However, this formula runs into two problems: first, x can be zero and the tangent is then undefined; second, the values of $\tan^{-1}$ do not span the entire range from 0 to 2π. This leads to a patchwork of conditions:

$$\theta = \begin{cases} \tan^{-1}\frac{y}{x}, & \text{if } x > 0\ , \\ \tan^{-1}\frac{y}{x} + \pi, & \text{if } x < 0, \\ \frac{\pi}{2}, & \text{if } x = 0 \text{ and } y \geq 0, \\ -\frac{\pi}{2}, & \text{if } x = 0 \text{ and } y < 0. \end{cases} \tag{7.14}$$

For the point at the origin ($x = 0, y = 0$) we obviously have $R = 0$, but the phase θ is not defined. Setting it to $\pi/2$ in equations (7.14) is for convenience.

7.9 Cosine of the Difference of Two Angles

We are already familiar with the Pythagoras identity and the complementary angle identity for trigonometric function. Below we dive deeper into the realm of various trigonometric identities. First, we derive a formula for the cosine of a difference of two angles.

Theorem 7.1 *The cosine of a difference of two angles is given by*

$$\cos(\alpha - \beta) = \cos\alpha\cos\beta + \sin\alpha\sin\beta. \tag{7.15}$$

Proof

Consider figure 7.7, where A, B, P, and Q are the end points of unit radii. The distance between points P and Q is the same as the distance between points A and B because points P and Q are obtained from points A and B by a rotation, which preserves distances. The coordinates of points P and Q are:

$$\begin{aligned} P &= (\cos\alpha, \sin\alpha), \\ Q &= (\cos\beta, \sin\beta). \end{aligned} \tag{7.16}$$

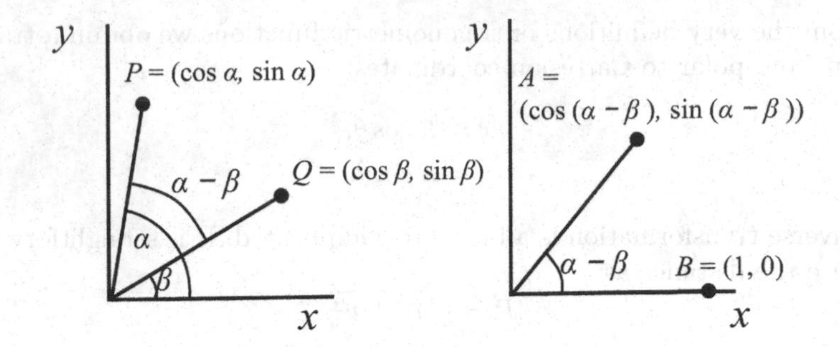

FIGURE 7.7
Cosine of a difference

Then the distance between these two points is

$$D_{PQ} = \sqrt{(\cos\alpha - \cos\beta)^2 + (\sin\alpha - \sin\beta)^2}. \tag{7.17}$$

The coordinates of points A and B are:

$$\begin{aligned} A &= (\cos(\alpha - \beta), \sin(\alpha - \beta)), \\ B &= (1, 0). \end{aligned} \tag{7.18}$$

Then the distance between these two points is

$$D_{AB} = \sqrt{(1 - \cos(\alpha - \beta))^2 + \sin^2(\alpha - \beta)}. \tag{7.19}$$

We know that these two distances are equal: $D_{AB} = D_{PQ}$. *Therefore*

$$\begin{aligned} &\sqrt{(\cos\alpha - \cos\beta)^2 + (\sin\alpha - \sin\beta)^2} = \\ &\sqrt{(1 - \cos(\alpha - \beta))^2 + \sin^2(\alpha - \beta)}. \end{aligned} \tag{7.20}$$

We square this equation to get:

$$\begin{aligned} &(\cos\alpha - \cos\beta)^2 + (\sin\alpha - \sin\beta)^2 = \\ &(1 - \cos(\alpha - \beta))^2 + \sin^2(\alpha - \beta). \end{aligned} \tag{7.21}$$

Next, we expand the squares:

$$\begin{aligned} &\cos^2\alpha + \cos^2\beta - 2\cos\alpha\cos\beta + \sin^2\alpha + \sin^2\beta - 2\sin\alpha\sin\beta = \\ &1 - 2\cos(\alpha - \beta) + \cos^2(\alpha - \beta) + \sin^2(\alpha - \beta). \end{aligned} \tag{7.22}$$

We use the Pythagoras identity (7.4) to replace every $\sin^2 + \cos^2$ *with* 1*:*

$$2 - 2\cos\alpha\cos\beta - 2\sin\alpha\sin\beta = 2 - 2\cos(\alpha - \beta). \tag{7.23}$$

We subtract 2 *from both sides and then divide by* -2*:*

$$\cos(\alpha - \beta) = \cos\alpha\cos\beta + \sin\alpha\sin\beta. \tag{7.24}$$

A formula for the cosine of the sum (rather than a difference) of two angles is as follows:

$$\cos(\theta + \beta) = \cos\theta\cos\beta - \sin\theta\sin\beta. \tag{7.25}$$

A proof of this formula is the subject of exercise 15 for this chapter.

Identities for the cosine of a sum and of a difference can be used to prove a recursive formula that expresses $\cos(n\theta)$ through cosine values of $\cos((n-1)\theta)$ and $\cos((n-2)\theta)$. This provides a practical algorithm for computing the value of $\cos(n\theta)$.

Theorem 7.2 *For any integer* n *and real* θ *it is the case that*

$$\cos(n\theta) = 2\cos\theta\cos((n-1)\theta) - \cos((n-2)\theta). \tag{7.26}$$

Proof

Consider two trigonometric identities that directly follow from the identities for the cosine of a sum or a difference:

$$\begin{aligned} \cos((n-1)\theta + \theta) &= \cos((n-1)\theta)\cos\theta - \sin((n-1)\theta)\sin\theta, \\ \cos((n-1)\theta - \theta) &= \cos((n-1)\theta)\cos\theta + \sin((n-1)\theta)\sin\theta. \end{aligned} \tag{7.27}$$

Adding them produces identity (7.26).

7.10 Back to the Identities for Complementary Angles

We use identity (7.15) to compute

$$\begin{aligned} &\cos\left(\frac{\pi}{2} - \theta\right) = \\ &\cos\frac{\pi}{2}\cos\theta + \sin\frac{\pi}{2}\sin\theta = \\ &\sin\theta. \end{aligned} \tag{7.28}$$

We have seen this equation before—it is identity (7.5) for complementary angles. Previously we proved this identity from figure 7.3, where the angle was in the first quadrant. This latest derivation extends this identity to all values of θ. Similarly,

$$\cos\beta = \sin\left(\frac{\pi}{2} - \beta\right). \tag{7.29}$$

A proof of this identity is the subject of exercise 17 for this chapter.

7.11 Sine of a Sum of Two Angles

Consider a sine of a sum of two angles: $\sin(\theta+\beta)$. Can we derive a formula for this expression?

Theorem 7.3 *The sine of a sum of two angles is given by*

$$\sin(\theta+\beta) = \sin\theta\cos\beta + \cos\theta\sin\beta. \tag{7.30}$$

Proof

We use equations for complementary angles (7.5) and theorem 7.1 for the cosine of a difference to get

$$\begin{aligned} \sin(\theta+\beta) &= \cos\left(\frac{\pi}{2}-\theta-\beta\right) \\ &= \cos\left(\frac{\pi}{2}-\theta\right)\cos\beta + \sin\left(\frac{\pi}{2}-\theta\right)\sin\beta \\ &= \sin\theta\cos\beta+\cos\theta\sin\beta. \end{aligned} \tag{7.31}$$

This is one of the most common trigonometric identities.

7.12 Sine and Cosine of a Double Angle

From the formulas for the sine and cosine of a sum of angles we get formulas for the double angle.

Theorem 7.4 *The sine and cosine of a double angle are given by*

$$\begin{aligned} \sin 2\theta &= 2\sin\theta\cos\theta, \\ \cos 2\theta &= \cos^2\theta - \sin^2\theta. \end{aligned} \tag{7.32}$$

Proof

Recall that

$$\begin{aligned} \sin(\theta+\beta) &= \sin\theta\cos\beta+\cos\theta\sin\beta, \\ \cos(\theta+\beta) &= \cos\theta\cos\beta-\sin\theta\sin\beta. \end{aligned} \tag{7.33}$$

We set $\theta=\beta$ to get the statement of the theorem.

In section 7.17 we will see another way to obtain these formulas.

7.13 One Way to Compute Trigonometric Functions

For most people, computing a numerical value of a sine or a cosine is a matter of entering the angle in a calculator and pressing a button. The calculator uses efficient and accurate algorithms for performing this calculation under the hood. Today these algorithms are based on calculus. However, the problem of relating lengths of the sides of triangles to the angles existed long before calculus was invented. During that time, people used lookup tables of chords.[2] The first such tables were computed by Hipparchus and then by Ptolemy in antiquity. Historians of mathematics reconstructed likely ways of how tables of chords were compiled. Here, we will not go into all the details of that ancient computation. Instead, we look at an illustration that uses some of the same ideas but is greatly simplified compared to what Hipparchus and Ptolemy did. We also will frame this computation in the modern terms of using trigonometric functions (rather than chords).

So, how do we compute the value of $\sin\theta$ without calculus? One possible way is to use the so-called small angle approximation, which states that if $|\theta|$ is small compared to 1, then $\sin\theta \approx \theta$. (Indeed, look at figure 7.3 and imagine that the angle is small. In that case, the opposite leg $\sin\theta$ of the right triangle becomes almost indistinguishable from the arc that is formed by angle θ, which would imply that $\sin\theta \approx \theta$.) Then from the Pythagoras identity it follows that for small angles $\cos\theta \approx 1$. We will use these approximations for computing the value of $\sin\theta$ for an arbitrary value of θ. We start from the trigonometric identity for the double angle:

$$\sin\theta = 2\sin\frac{\theta}{2}\cos\frac{\theta}{2}. \tag{7.34}$$

The argument of the trigonometric functions in the right-hand side is smaller by the absolute value than that in the left-hand side. We can repeat this trick to halve the argument again:

$$\begin{aligned}\sin\theta &= 2\sin\frac{\theta}{2}\cos\frac{\theta}{2}\\ &= 4\sin\frac{\theta}{4}\cos\frac{\theta}{4}\left(\cos^2\frac{\theta}{4} - \sin^2\frac{\theta}{4}\right).\end{aligned} \tag{7.35}$$

We can continue this process indefinitely. The absolute value of the argument of trigonometric functions in the right-hand side will be getting progressively smaller, at the cost of the expression getting longer. At some point the argument will become small enough for the small angle approximation to work well. We just have to preset some small tolerance value d, and then use the small angle approximation when the argument of all trigonometric functions drops below d, replacing then all sines with the value of the argument, and all cosines with 1. The accuracy of the computation will then depend on d: for smaller values of d the computation becomes longer and more tedious, but also more accurate because the small angle approximation works better. Table 7.1 shows results for computing the sine of 0.5 using this algorithm. For comparison, the exact value is given by $\sin 0.5 \approx 0.479425538604203$.

This is a very labor-intensive way to compute a value of trigonometric functions, and modern algorithms are much more economical. But in the absence of modern methods this is one way to compute $\sin\theta$. In antiquity they

[2] A chord is a line segment connecting two points on a circle. The length of a chord is defined by the radius of the circle and the angle θ of the circular arc that is formed by the

TABLE 7.1
Approximation of the sine function.

Tolerance d	**Approximate value for** $\sin 0.5$
0.1	0.48638150095939636
0.01	0.48035385862459756
0.001	0.4795424605661942
0.0001	0.47943285356824394
0.00001	0.4794264530282436

computed chord tables using methods that share some of the ideas presented in this section, but are even more cumbersome.

7.14 More Trigonometric Identities

The trigonometric identities for the sums of angles and the Pythagoras identity are the basis for more identities, which are presented in this section.

7.14.1 Tangent of a Sum of Angles

Theorem 7.5 *The tangent of a sum of angles is given by*

$$\tan(\theta+\beta) = \frac{\tan\theta + \tan\beta}{1-\tan\theta\tan\beta}. \tag{7.36}$$

Proof

We use trigonometric identities for sine and cosine to get:

$$\begin{aligned}
\tan(\theta+\beta) &= \frac{\sin(\theta+\beta)}{\cos(\theta+\beta)} \\
&= \frac{\sin\theta\cos\beta + \cos\theta\sin\beta}{\cos\theta\cos\beta - \sin\theta\sin\beta} \\
&= \frac{\left(\dfrac{\sin\theta\cos\beta + \cos\theta\sin\beta}{\cos\theta\cos\beta}\right)}{\left(\dfrac{\cos\theta\cos\beta - \sin\theta\sin\beta}{\cos\theta\cos\beta}\right)} \\
&= \frac{\dfrac{\sin\theta\cos\beta}{\cos\theta\cos\beta} + \dfrac{\cos\theta\sin\beta}{\cos\theta\cos\beta}}{\dfrac{\cos\theta\cos\beta}{\cos\theta\cos\beta} - \dfrac{\sin\theta\sin\beta}{\cos\theta\cos\beta}} \\
&= \frac{\tan\theta + \tan\beta}{1-\tan\theta\tan\beta}.
\end{aligned} \tag{7.37}$$

end points of the chord. For a unit circle, the length of a chord is given by $2\sin\theta/2$.

7.14.2 Sine of a Half-Angle

Theorem 7.6 *The sine of a half-angle is given by*

$$\sin\frac{\beta}{2} = \pm\sqrt{\frac{1-\cos\beta}{2}}. \tag{7.38}$$

Here the sign should be chosen using the correct quadrant for angle $\beta/2$.

Proof

From the cosine of a double angle and from the Pythagoras identity, we get:

$$\begin{aligned}\cos 2\theta &= \cos^2\theta - \sin^2\theta \\ &= 1 - 2\sin^2\theta.\end{aligned} \tag{7.39}$$

We solve this equation for $\sin\theta$ *to get*

$$\sin\theta = \pm\sqrt{\frac{1-\cos 2\theta}{2}}. \tag{7.40}$$

Here the sign should be chosen using the correct quadrant for the angle θ. We introduce a new notation $\beta = 2\theta$ to get the statement of the theorem.

7.14.3 Cosine of a Half-Angle

Theorem 7.7 *The cosine of a half-angle is given by*

$$\cos\frac{\beta}{2} = \pm\sqrt{\frac{1+\cos\beta}{2}}. \tag{7.41}$$

Here the sign should be chosen using the correct quadrant for angle $\beta/2$.

A proof of this theorem is the subject of exercise 18 for this chapter.

7.14.4 Sums and Differences of Trigonometric Functions

Another group of trigonometric identities deals with sums or differences of trigonometric functions.

Theorem 7.8 *A sum of cosines can be represented as*

$$\cos\theta + \cos\beta = 2\cos\frac{\theta+\beta}{2}\cos\frac{\theta-\beta}{2}. \tag{7.42}$$

Proof

We use theorem 2.19 to represent θ and β through the sum and the difference of $u = (\theta+\beta)/2$ and $v = (\theta-\beta)/2$ and then apply the identities for the cosine of a sum and the cosine of a difference:

$$\begin{aligned}\cos\theta + \cos\beta &= \cos(u+v) + \cos(u-v) \\ &= \cos u\cos v - \sin u\sin v + \cos u\cos v + \sin u\sin v \\ &= 2\cos u\cos v \\ &= 2\cos\frac{\theta+\beta}{2}\cos\frac{\theta-\beta}{2}.\end{aligned} \tag{7.43}$$

Theorem 7.9 *A difference of cosines can be represented as*

$$\cos\theta - \cos\beta = -2\sin\frac{\theta+\beta}{2}\sin\frac{\theta-\beta}{2}. \tag{7.44}$$

Proof

This identity is derived analogously:

$$\begin{aligned}\cos\theta - \cos\beta &= \cos(u+v) - \cos(u-v)\\ &= \cos u\cos v - \sin u\sin v - \cos u\cos v - \sin u\sin v\\ &= -2\sin u\sin v\\ &= -2\sin\frac{\theta+\beta}{2}\sin\frac{\theta-\beta}{2}.\end{aligned} \tag{7.45}$$

Since the sine function is odd, we can wrap the minus sign into the argument of one of the sines:

$$\begin{aligned}\cos\theta - \cos\beta &= -2\sin\frac{\theta+\beta}{2}\sin\frac{\theta-\beta}{2}\\ &= 2\sin\frac{\theta+\beta}{2}\sin\frac{\beta-\theta}{2}.\end{aligned} \tag{7.46}$$

Similarly, we get for the sum of sines:

Theorem 7.10 *The sum of sines can be represented as*

$$\sin\theta + \sin\beta = 2\sin\frac{\theta+\beta}{2}\cos\frac{\theta-\beta}{2}. \tag{7.47}$$

Proof

$$\begin{aligned}\sin\theta + \sin\beta &= \sin(u+v) + \sin(u-v)\\ &= \sin u\cos v + \cos u\sin v + \sin u\cos v - \cos u\sin v\\ &= 2\sin u\cos v\\ &= 2\sin\frac{\theta+\beta}{2}\cos\frac{\theta-\beta}{2}.\end{aligned} \tag{7.48}$$

The final identity in this group is for the difference of sines.

Theorem 7.11 *The difference of sines is computed as follows.*

$$\sin\theta - \sin\beta = 2\cos\frac{\theta+\beta}{2}\sin\frac{\theta-\beta}{2}. \tag{7.49}$$

A proof of this theorem is the subject of exercise 19 for this chapter.

7.14.5 Products of Trigonometric Functions

The above identities express sums of trigonometric functions through a product. We can flip them to express a product through a sum. We again recall theorem 2.19 and the transition between variables (u, v) and (θ, β):

$$\begin{aligned} u + v &= \frac{\theta + \beta}{2} + \frac{\theta - \beta}{2} = \theta, \\ u - v &= \frac{\theta + \beta}{2} - \frac{\theta - \beta}{2} = \beta. \end{aligned} \tag{7.50}$$

Then, we get from theorem 7.11 the following:

$$\cos u \sin v = \frac{1}{2}\left(\sin(u+v) - \sin(u-v)\right). \tag{7.51}$$

This and three other similar identities are given by the following theorem:

Theorem 7.12 *Products of sines and cosines can be represented as*

$$\begin{aligned} \cos u \sin v &= \frac{1}{2}\left(\sin(u+v) - \sin(u-v)\right), \\ \cos u \cos v &= \frac{1}{2}\left(\cos(u+v) + \cos(u-v)\right), \\ \sin u \sin v &= \frac{1}{2}\left(\cos(u-v) - \cos(u+v)\right), \\ \sin u \cos v &= \frac{1}{2}\left(\sin(u+v) + \sin(u-v)\right). \end{aligned} \tag{7.52}$$

A proof of this theorem is the subject of exercise 20 for this chapter.

We already looked at a simplified mathematical model for most radio signals that is given by $E\sin(2\pi ft)$, where t is time, and f is called the frequency of the signal (p. 156). This model describes an oscillating electromagnetic wave. Engineers who design electronics for telecommunications equipment face a problem: some technical requirements (including the antenna size) favor using higher values of frequency f, but other requirements (such as processing the signal by the receiver) favor lower values of f. To reconcile these conflicting requirements, engineers use a well-known trick: the incoming high-frequency signal is multiplied by another signal that is internally generated by the receiver and that uses some other frequency f_1. Mathematically, this is expressed as $E\sin(2\pi ft)\cdot\cos(2\pi f_1 t)$. According to theorem 7.12,

$$E\sin(2\pi ft)\cdot\cos(2\pi f_1 t) = \frac{E}{2}(\sin(2\pi(f+f_1)t) + \sin(2\pi(f-f_1)t)). \tag{7.53}$$

We see that this operation produces two signals, one at the sum $(f + f_1)$ and another at the difference $(f - f_1)$ of frequencies f and f_1. If $f \approx f_1$, the second signal has a much lower frequency and is much easier to process. This makes it possible to transmit a high frequency signal, perform signal mixing according to equation (7.53), and then do the bulk of the processing on a low frequency signal, making the equipment cheaper and more compact. (The extraneous signal at the sum of the two frequencies is easy to suppress.)

This is probably by far the most frequent application of trigonometric identities as measured by instances of use, as it is employed by billions of cell phones thousands of times each second!

7.15 Multiplication of Complex Numbers

In section 2.7 we learned that a complex number $z = x + iy$ can be viewed as a point on the (x, y) plane. Let's see what happens if a complex number is represented in polar rather than Cartesian coordinates. Then

$$z = R\cos\phi + iR\sin\phi, \tag{7.54}$$

where R is the modulus of z, and the angle ϕ is called the phase. Below we consider the effect that multiplication has on the polar coordinates of two complex numbers, $z_1 = x_1 + iy_1$ and $z_2 = x_2 + iy_2$. We already know that (see equation (2.83))

$$z_1 z_2 = (x_1 x_2 - y_1 y_2) + i(x_1 y_2 + x_2 y_1). \tag{7.55}$$

The transition between polar and Cartesian coordinates is given in section 7.8. We substitute $x_1 = R_1\cos\phi_1; y_1 = R_1\sin\phi_1; x_2 = R_2\cos\phi_2; y_2 = R_2\sin\phi_2$ in equation (7.55) to get:

$$\begin{aligned} z_1 z_2 &= R_1 R_2(\cos\phi_1\cos\phi_2 - \sin\phi_1\sin\phi_2) \\ &\quad + iR_1 R_2(\cos\phi_1\sin\phi_2 + \sin\phi_1\cos\phi_2). \end{aligned} \tag{7.56}$$

In the right-hand side, we recognize formulas for the cosine of a sum (7.25) and the sine of a sum (7.30) of two angles. Then

$$z_1 z_2 = R_1 R_2(\cos(\phi_1 + \phi_2) + i\sin(\phi_1 + \phi_2)). \tag{7.57}$$

Amazingly, the product of two complex numbers is defined by the following two rules:

1. The modulus of the product is the product of the moduli of the factors: $R = R_1 R_2$.

2. The phase of the product is the sum of the phases of the factors: $\phi = \phi_1 + \phi_2$.

> The last statement means that if we take a complex number z_1 and multiply it by some other number z_2, in polar coordinates the point that corresponds to number z_1 is rotated by the phase of number z_2—that is, equation (7.57) links multiplication of complex numbers to rotation. On p. 52 we stated that a further generalization of complex numbers to three and more dimensions loses some useful properties, such as commutativity of the product. In three and higher dimensions, rotations are not commutative (see p. 154). That is the primary reason for the lack of generalizations of complex numbers that would preserve all their properties, in particular, commutativity of multiplication.

In polar coordinates the ratio of two complex numbers is computed more elegantly than what is given by equation (2.98). For $z_1 = R_1(\cos\phi_1 + i\sin\phi_1)$ and $z_2 = R_2(\cos\phi_2 + i\sin\phi_2)$ we get

$$\frac{z_1}{z_2} = \frac{R_1}{R_2}(\cos(\phi_1 - \phi_2) + i\sin(\phi_1 - \phi_2)). \tag{7.58}$$

From the formula for multiplying complex numbers we obtain de Moivre's theorem:

Theorem 7.13 *For any real number ϕ and integer n it is the case that*

$$(\cos\phi + i\sin\phi)^n = \cos n\phi + i\sin n\phi. \tag{7.59}$$

Proof

The proof is by induction:

1. *For $n = 1$ the statement of the theorem holds.*
2. *We assume that the theorem is true for n and consider it for $n+1$. We use*

$$(\cos\phi + i\sin\phi)^{n+1} = (\cos\phi + i\sin\phi)^n \cdot (\cos\phi + i\sin\phi) \tag{7.60}$$

and apply the theorem for n to get

$$(\cos\phi + i\sin\phi)^{n+1} = (\cos n\phi + i\sin n\phi) \cdot (\cos\phi + i\sin\phi). \tag{7.61}$$

Then the desired result

$$(\cos\phi + i\sin\phi)^{n+1} = \cos(n+1)\phi + i\sin(n+1)\phi \tag{7.62}$$

directly follows from equation (7.57).

A real number can be viewed as a particular case of a complex number whose imaginary part is zero. For a negative real number we must have $\cos\phi < 0$ and $\sin\phi = 0$, which implies $\phi = \pi$. Per equation (7.57), the phase of a product of two negative real numbers will then be equal to $\pi + \pi = 2\pi$, which is the same as having a zero phase—that is, corresponding to a positive real number. This is consistent with the fact that a product of two negative real numbers is positive.

7.16 Back to the Fundamental Theorem of Algebra

All the things that we have already learned about complex numbers make a good foundation to discuss the Fundamental Theorem of Algebra again. In this section we outline a proof of this theorem. This will still not be a rigorous proof because we will use some notions that we have not introduced, such as function continuity. With some effort, the holes in the argument below can be filled to get an elegant and complete proof.

As a reminder, we discussed the Fundamental Theorem of Algebra in section 5.8. It states that every nonzero polynomial with degree $n \geq 1$ has at least one complex root. Here, we consider a polynomial of n-th degree:

$$P(z) = a_0 + \ldots + a_{n-1}z^{n-1} + a_n z^n. \tag{7.63}$$

Note that we allow the argument (and coefficients) to be complex. This is necessary because we are looking for a proof that there exists a complex root for this polynomial.

Next, we consider two cases. In the first case, $a_0 = 0$. Then, we can check by a direct substitution that $z = 0$ is a root of this polynomial.

In the second case, $a_0 \neq 0$. By the direct substitution of $z = 0$, we can see that the zero value is now *not* a root of the polynomial. Here is an argument that shows the existence of at least one root at some point $z \neq 0$.

We start from a thought experiment. Let's think what happens with the value of a complex number if we "walk" in the complex plane counterclockwise in a circle with the center at the origin. The modulus R of the complex number $z = R(\cos\phi + i\sin\phi)$ will remain constant. The phase ϕ will increase as we move along the circle. After we have completed one full revolution, the phase will have increased by 2π. Since the walk ends up at the starting point, the value of the complex number at the end must be the same as the value at the start. Indeed, this is true because the trigonometric functions in the expression $z = R(\cos\phi + i\sin\phi)$ are periodic with the period of 2π.

What happens with the modulus and phase of z^n if we "walk" along the same circle? In that case, the modulus will again remain constant, but de

Moivre's theorem predicts that the phase will increment by $2\pi n$. The end value of z^n will still match the start value because now the arguments of the trigonometric functions will increase by n full periods.

Let's now extend this thought experiment and think what happens with the value of polynomial $P(z)$ if we take the same walk. Since we end up at the starting point, the value of the polynomial must end up at the starting value, regardless of the radius of the circle. In other words, equation $z_{\text{end}} = z_{\text{start}}$ implies that

$$P(z_{\text{end}}) = P(z_{\text{start}}). \tag{7.64}$$

Similarly to the case of z or z^n, the phase of $P(z)$ can vary along the circle, but since we end up at the starting point, the phases at the start and the end of a "walk" must differ by some integer multiple of 2π. This multiple, which we denote k, can be equal to zero or have any other integer value, but it cannot be noninteger because then the condition $P(z_{\text{end}}) = P(z_{\text{start}})$ would not hold. Now let's look into possible values of k for circles of different radii.

For a very large circle, the last term $a_n z^n$ in the polynomial makes by far the largest contribution to the total value. Therefore, for a large enough circle the change in the phase of the polynomial along a circular walk must be driven by the highest-order term. This means that the phase will be incremented by $2\pi n$. Note that even though lower-order terms may affect the values of the phase of $P(z)$ at different points on the circle, at the end point it must increase exactly by $2\pi n$ as compared to the starting point.

For a very small circle, the first term a_0 in the polynomial makes by far the largest contribution to the total value. This means that the phase will not be incremented at all at the end of the circle as compared to the starting point. Similarly, even though higher-order terms may affect the variation of the phase of $P(z)$ along the circle, at the end its value must get exactly to the starting point.

For a circle with a radius in between the large and the small one, the phase of $P(z)$ must be incremented by an integer multiple of 2π, but not necessarily by $2\pi n$ or by zero.

Now let's again consider a very large circle and then think what happens if we continuously shrink its radius until it becomes very small. We expect the total increment of the phase of $P(z)$ for a "walk" along a circle to vary continuously as a function of the radius of the circle. However, this statement runs into a problem: as we shrink the circle from a very large to a very small one, the phase increment must somehow change from $2\pi n$ to zero, while remaining an integer multiple of 2π at every value of the radius. This requires it to have one or more discrete jumps in value by a multiple of 2π. Such discrete jumps in the phase increment are clearly in contradiction with the assumption that all variables vary smoothly and continuously. How can that be?

The resolution of this problem is possible due to the fact that the phase is *not defined* for the complex number that is equal to zero (see p. 161). Now consider a circle that passes through a point where $P(z) = 0$. The phase of $P(z)$ at that point is not defined, and the total increment of the phase along

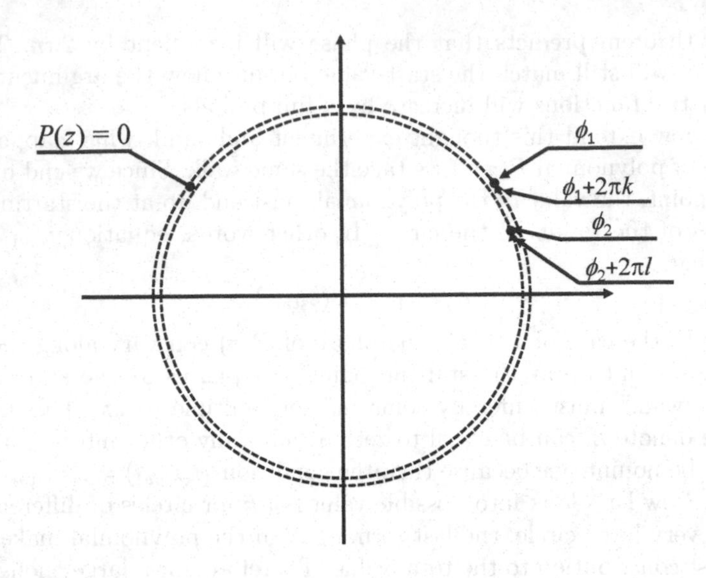

FIGURE 7.8
Phase increases along two circles

the circle passing through that point is not defined either. For a circle of a slightly larger radius, the total phase increment may be equal to $2\pi k$, and for a circle of a slightly smaller radius, the total phase increment may be equal to $2\pi l$, where k, l are some integers (see figure 7.8). Since the phase increment is not defined for the circle that passes through a point where $P(z) = 0$, the continuity requirement $k = l$ falls apart and no longer contradicts the assumption that the total phase increment must take discrete values.

Therefore, we must have at least one point where $P(z) = 0$ to ensure a transition from the total phase increment of $2\pi n$ to zero for a shrinking circle, which means that $P(z)$ must have a complex root!

7.17 Euler's Formula

Much of the material in this section is given without proof. It is included because it is cool and has important practical applications.

Here, we explore the exponential function for a complex argument. A proper definition of a complex exponent is beyond the scope of this book, as is the proof of the key formula, which bears the name of Leonhard Euler. We just state that

$$e^{i\theta} = \cos\theta + i\sin\theta. \tag{7.65}$$

This jewel of mathematics has countless applications. It is also an amazing example of linking two seemingly distant mathematical functions: turns out, the exponent and the trigonometric functions are cousins! We also can represent trigonometric functions through the exponent:

$$\begin{aligned}\sin\theta &= \frac{e^{i\theta} - e^{-i\theta}}{2i},\\ \cos\theta &= \frac{e^{i\theta} + e^{-i\theta}}{2}.\end{aligned} \tag{7.66}$$

These formulas are proved by substituting Euler's formula in the right-hand side. The modulus of the both sides of equation (7.65) is equal to 1. Indeed,

$$|\cos\theta + i\sin\theta| = \sqrt{\cos^2\theta + \sin^2\theta} = 1. \tag{7.67}$$

Euler's formula also yields a simple derivation for several trigonometric identities. For example, consider the complex exponent of a double argument:

$$e^{i2\theta} = \cos 2\theta + i\sin 2\theta. \tag{7.68}$$

Note that the left-hand side is the square of $e^{i\theta}$. This means that we can compute it by squaring the original Euler's formula (7.65):

$$\begin{aligned}e^{i2\theta} &= (\cos\theta + i\sin\theta)^2\\ &= \cos^2\theta - \sin^2\theta + 2i\cos\theta\sin\theta.\end{aligned} \tag{7.69}$$

Comparing the right-hand sides of equations (7.68) and (7.69) yields

$$\cos 2\theta + i\sin 2\theta = \cos^2\theta - \sin^2\theta + 2i\cos\theta\sin\theta. \tag{7.70}$$

Since the real and imaginary parts must be pairwise equal, we get:

$$\begin{aligned}\cos 2\theta &= \cos^2\theta - \sin^2\theta,\\ \sin 2\theta &= 2\cos\theta\sin\theta.\end{aligned} \tag{7.71}$$

These are the familiar double-angle formulas. We can extend this idea and get a formula for the sine and cosine of a multiple of an angle. We start from

$$\begin{aligned}e^{in\theta} &= \cos n\theta + i\sin n\theta\\ &= (\cos\theta + i\sin\theta)^n.\end{aligned} \tag{7.72}$$

The value of $(\cos\theta + i\sin\theta)^n$ is computed by using the binomial expansion (see section 5.2). In the expanded result, the terms with even powers of i will be real, and the terms with odd powers of i will be imaginary. Next, we equate $\cos n\theta$ to the real terms in the expansion, and $i\sin n\theta$ to the imaginary terms.

This produces the following general formulas for the trigonometric functions of a multiple of an angle. After a bulky derivation, the final result is as follows:

$$\begin{aligned} \sin n\theta &= \sum_{k=0}^{n} \binom{n}{k} \cos^k \theta \sin^{n-k} \theta \sin \frac{(n-k)\pi}{2}, \\ \cos n\theta &= \sum_{k=0}^{n} \binom{n}{k} \cos^k \theta \sin^{n-k} \theta \cos \frac{(n-k)\pi}{2}. \end{aligned} \tag{7.73}$$

Here $\sin(n-k)\pi/2$ and $\sin(n-k)\pi/2$ take values of -1, 0, and 1. Their role is to set the correct sign for each term in the sum or to null that term.

Euler's formula raises the number e to a complex power. It is also possible to raise a complex number to a complex power. Without going into all the details of complex exponentiation, here is an interesting fact: the value of $i^i \approx 0.20787957635$ is a real number! Indeed, let's start from Euler's formula (7.65) and set $\theta = \pi/2$ there. Since $\cos \pi/2 = 0$ and $\sin \pi/2 = 1$, we get

$$e^{i\pi/2} = i. \tag{7.74}$$

We raise this equation to the power of i (omitting the mechanics of this procedure):

$$\left(e^{i\pi/2}\right)^i = i^i. \tag{7.75}$$

In the left-hand side, we get

$$\begin{aligned} \left(e^{i\pi/2}\right)^i &= e^{i^2\pi/2} \\ &= e^{-\pi/2} \\ &\approx 0.20787957635. \end{aligned} \tag{7.76}$$

Using Euler's formula, a complex number can be represented as

$$z = R(\cos\theta + i\sin\theta) = Re^{i\theta}. \tag{7.77}$$

Then a product of two complex numbers is given by

$$z_1 z_2 = R_1 R_2 e^{i(\theta_1+\theta_2)}, \tag{7.78}$$

which reiterates equation (7.57). It also yields a straightforward proof of theorem 2.25, which states that $|z_1 z_2| = |z_1||z_2|$.

7.18 Three Trigonometric Inequalities

In this section we prove three curious trigonometric inequalities. Their proofs feel a bit different from the proofs of trigonometric identities earlier in this chapter. One proof uses mathematical induction, which is less common in trigonometry.

Theorem 7.14 *For any $0 \le p \le \pi; 0 \le q \le \pi$ it is the case that*

$$\sqrt{\sin p \sin q} \le \sin \frac{p+q}{2}. \tag{7.79}$$

Proof

For these values of p, q the values of the sine function are positive. We use the fact that the arithmetic mean is greater or equal to the geometric mean (theorem 3.17):

$$\sqrt{\sin p \sin q} \le \frac{\sin p + \sin q}{2}. \tag{7.80}$$

We apply the identity for the sum of sines (theorem 7.10):

$$\sqrt{\sin p \sin q} \le \sin \frac{p+q}{2} \cos \frac{p-q}{2}. \tag{7.81}$$

Since the cosine in the right-hand side does not exceed 1, we get the desired inequality:

$$\sqrt{\sin p \sin q} \le \sin \frac{p+q}{2}. \tag{7.82}$$

Note that for small values of p and q, the small angle approximation (see p. 165) reduces inequality (7.79) to

$$\sqrt{pq} \le \frac{p+q}{2}, \tag{7.83}$$

which is the inequality for the arithmetic and geometric means (theorem 3.17). However, theorem 7.14 works for p, q ranging from 0 to π, and not just for small values.

Theorem 7.15 *For any p it is the case that*

$$\sin^4 p + \cos^4 p \ge \frac{1}{2}. \tag{7.84}$$

Below are two proofs of this inequality.

Proof

We recall the Cauchy-Schwarz inequality (see equation (3.38)):

$$\left(\sum_{n=1}^{N} x_n y_n\right)^2 \le \left(\sum_{n=1}^{N} x_n^2\right) \cdot \left(\sum_{n=1}^{N} y_n^2\right). \tag{7.85}$$

We set $N = 2; x_1 = x_2 = 1; y_1 = \sin^2 p; y_2 = \cos^2 p$. *Then*

$$(\sin^2 p + \cos^2 p)^2 \leq (1+1) \cdot (\sin^4 p + \cos^4 p). \tag{7.86}$$

The left-hand side is equal to 1 *because of the Pythagoras identity, and* $(1+1)$ *in the right-hand side simplifies to* 2. *This yields the statement of the theorem.*

Proof

We start from $|\sin 2p| \leq 1$ *and use the formula for the sine of a double angle (theorem 7.4) to get*

$$|2 \sin p \cos p| \leq 1. \tag{7.87}$$

We square this inequality and divide the result by 2 *to get*

$$2 \sin^2 p \cos^2 p \leq \frac{1}{2}. \tag{7.88}$$

The left-hand side is as follows:

$$(\sin^2 p + \cos^2 p)^2 - (\sin^4 p + \cos^4 p) \leq \frac{1}{2}. \tag{7.89}$$

(Indeed, if we expand the left-hand side in equation (7.89), we will get the left-hand side of equation (7.88).) The first term in the left-hand side of equation equation (7.89) is equal to 1 *because of the Pythagoras identity. Then*

$$1 - (\sin^4 p + \cos^4 p) \leq \frac{1}{2}. \tag{7.90}$$

From here, we shuffle the terms to produce the statement of the theorem.

Theorem 7.16 *For any natural* n *and real* θ *it is the case that*

$$|\sin n\theta| \leq n|\sin \theta|. \tag{7.91}$$

Proof

We use mathematical induction:

1. *For* $n = 1$ *this inequality is true.*
2. *We assume that it is true for* n *and consider the case of* $n+1$. *The formula for the sine of a sum yields:*

$$|\sin(n+1)\theta| = |\sin n\theta \cos\theta + \cos n\theta \sin\theta|. \tag{7.92}$$

We use theorem 3.11 to split the right-hand side:

$$\begin{aligned} |\sin(n+1)\theta| &= |\sin n\theta \cos\theta + \cos n\theta \sin\theta| \\ &\leq |\sin n\theta \cos\theta| + |\cos n\theta \sin\theta| \\ &= |\sin n\theta||\cos\theta| + |\cos n\theta||\sin\theta|. \end{aligned} \tag{7.93}$$

Since absolute values of the cosine function do not exceed 1,

$$|\sin(n+1)\theta| \leq |\sin n\theta| + |\sin\theta|. \tag{7.94}$$

The statement of the theorem is true for n*. Therefore, the previous inequality produces*

$$|\sin(n+1)\theta| \leq n|\sin\theta| + |\sin\theta|. \tag{7.95}$$

We factor out the sine function in the right-hand side to get the desired inequality for $n+1$*:*

$$|\sin(n+1)\theta| \leq (n+1)|\sin\theta|. \tag{7.96}$$

Exercise 28 for this chapter shows that this inequality cannot be generalized to the case of noninteger values of n. Another proof of theorem 7.16 is the subject of exercise 28 for chapter 9.

7.19 Analytical Geometry

Trigonometry is a great help when we want to solve a geometric problem using algebraic means. Here, we prove two important theorems that are often used in applications, the law of cosines and the law of sines. Both link the lengths of the sides of triangles with the angles.

Theorem 7.17 *(The Law of Cosines) The length of a side of a triangle is expressed through the two other sides and the angle between them as follows:*

$$c^2 = a^2 - 2ab\cos\theta + b^2, \tag{7.97}$$

where notations are defined in figure 7.9.

Proof

Consider the distance between points A *and* B*. The coordinates of* A *are given by* $(b\cos\theta, b\sin\theta)$*. The coordinates of* B *are given by* $(a, 0)$*. Then*

$$c^2 = (a - b\cos\theta)^2 + (0 - b\sin\theta)^2. \tag{7.98}$$

We expand the squares and simplify:

$$c^2 = a^2 - 2ab\cos\theta + b^2\cos^2\theta + b^2\sin^2\theta \tag{7.99}$$

or

$$c^2 = a^2 - 2ab\cos\theta + b^2. \tag{7.100}$$

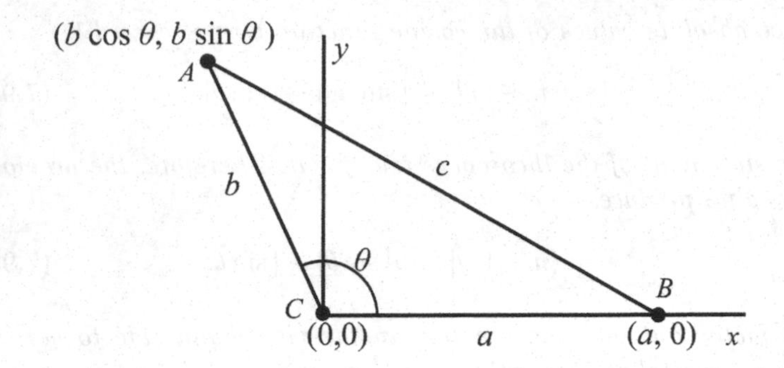

FIGURE 7.9
Law of cosines

The law of cosines allows us to compute a side of a triangle from the angle and the two other sides, which comes very handy in many applications. The Law of Sines is another useful theorem; it links the lengths of sides and two angles:

Theorem 7.18 *Two sides and adjacent angles of a triangle are linked by the following expression:*

$$c \sin \beta = b \sin \gamma, \tag{7.101}$$

where notations are defined in figure 7.10.

Proof
In a triangle we drop a perpendicular from the vertex that is common to sides b and c, which produces two right triangles. Let's compute the length of that

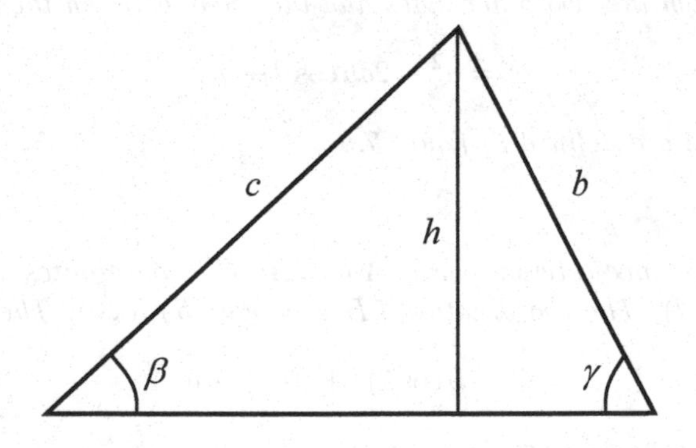

FIGURE 7.10
Law of sines

perpendicular from these two triangles:

$$
\begin{aligned}
h &= c \sin \beta, \\
h &= b \sin \gamma.
\end{aligned}
\tag{7.102}
$$

Equating these two expressions for h yields the statement of the theorem.

In 1735 about 20 French and Spanish scientists embarked on the expedition to measure the length of the Earth meridian and to accurately determine the shape of the Earth. To do that, they used the device that is called quadrant, which is essentially a telescope mounted on a precise angle measurement platform. At every step, they positioned three wooden poles on the terrain to form a triangle. Knowing the length of one side of the triangle and using quadrants to measure angles between the poles, they determined the lengths of the other two sides of the triangle. (This procedure is known as triangulation.) Then they used these two newly measured sides as new baselines to set up and measure new triangles. Of course, this whole procedure would be impossible without trigonometry.

The next step was to measure the angle to known stars from selected vertices of the triangulation mesh, which gave them the latitude of these points. Combining the data about the lengths and latitudes, they were able to compute the length of the Earth meridian and to establish that the Earth is slightly oblate (flattened at the poles).

This expedition lasted for many years and had a great effect on the long-range navigation. It also showed that a team of scientists working together for many years can produce a breakthrough. Many modern scientific projects do require international cooperation of large teams over years.

Precise geodetic measurements have gone a long way since the 18th century. Today most navigation instruments, such as GPS receivers, use the WGS84 reference system for the Earth shape. This system requires periodic maintenance because tectonic plates constantly drift. While the GPS technology vastly improved the speed and accuracy of positioning, its mathematical foundation is conceptually similar to what scientists did in the 18th century—that is, it also uses triangulation.

Exercises

1. Use equations (7.5) to prove that

$$
\begin{aligned}
\sin(\pi - \theta) &= \sin \theta, \\
\cos(\pi - \theta) &= -\cos \theta.
\end{aligned}
\tag{7.103}
$$

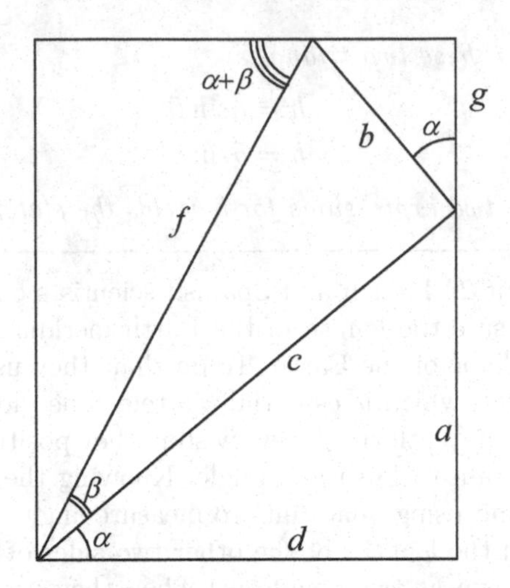

FIGURE 7.11
Formulas for the sine and cosine of a sum

2. Prove that if the sine function is not algebraic (see p. 155), then the cosine function is not algebraic.
3. Prove that $zz^* = |z|^2$ using z in the polar form.
4. Prove that
$$\sum_{n=0}^{N} \binom{N}{n} \sin^{2n}\theta \cos^{2N-2n}\theta = 1. \tag{7.104}$$
(Hint: note the similarity of the expression in the right-hand side with that for the binomial expansion.)
5. Consider the Law of Cosines (theorem 7.17) for a right triangle. In that case, $\theta = \pi/2$ and $\cos\theta = 0$, which produces $c^2 = a^2 + b^2$. While this is the statement of the Pythagoras theorem, the proof of the Law of Cosines on p. 179 cannot be used to prove the Pythagoras theorem. Explain why.
6. Obtain formulas for the sine and cosine of a sum from figure 7.11.
7. Determine the intervals of θ where $\sin\theta$ and $\cos\theta$ are monotonic and prove that equations (7.7) and (7.8) do enumerate all solutions of respective equations $\sin\theta = a$ and $\cos\theta = a$.
8. Prove that $\cos(n\theta)$ can be expressed as a polynomial of $\cos\theta$. (This polynomial is known as the Chebyshev polynomial of the first kind.) Use two methods:

   (a) By induction, using identity (7.26) or one of the identities in equations (7.52).

(b) Using de Moivre's theorem 7.13. (Hint: think what happens if the left-hand side of equation (7.59) is expanded. Which powers of $\sin\theta$ would contribute to the real part of that expression?)

9. Prove that equation (7.41) can be obtained from identity (7.26) by using a specific value of n there.

10. Prove that for $0 \leq x \leq 1, 0 \leq y \leq 1$ it is the case that

$$\sin^{-1} x + \sin^{-1} y = \sin^{-1}\left(x\sqrt{1-y^2} + y\sqrt{1-x^2}\right). \tag{7.105}$$

11. Prove that

$$\cos 3\theta = 4\cos^3\theta - 3\cos\theta. \tag{7.106}$$

12. Use $\cos\pi/3 = 1/2; \sin\pi/3 = \sqrt{3}/2$, and equation (7.106) to prove that

$$\cos 3\theta = 4\cos\theta\cos\left(\frac{\pi}{3} - \theta\right)\cos\left(\frac{\pi}{3} + \theta\right). \tag{7.107}$$

13. Prove that for all $\theta \neq (2n+1)\pi$ it is the case that

$$\sqrt{\frac{1-\sin\theta}{1+\sin\theta}} = \left|\frac{1-\tan\frac{\theta}{2}}{1+\tan\frac{\theta}{2}}\right|. \tag{7.108}$$

(Hint: express both occurrences of the 1 in the left-hand side through $\sin\theta/2, \cos\theta/2$ by using the Pythagorean identity.)

14. Prove that for any real value of t,

$$\tan^{-1} t + \tan^{-1}\frac{1}{t} = \frac{\pi}{2}. \tag{7.109}$$

15. Prove equation (7.25) directly from equation (7.15).

16. Modify figure 7.7 to prove the identity for the cosine of a sum using the same logic as in the proof of theorem 7.1.

17. Prove identity (7.29) from equation (7.28).

18. Use the proof of theorem 7.6 as a template to prove theorem 7.7.

19. Prove theorem 7.11 using proofs of theorems 7.10 and 7.9 as a template.

20. Prove the last three equations in (7.52).

21. Substitute $z_1 = R_1(\cos\phi_1 + i\sin\phi_1), z_2 = R_2(\cos\phi_2 + i\sin\phi_2)$ in equation (2.98) and obtain equation (7.58).

22. Use equation (7.65) to derive identities for the sine and cosine of a sum and a difference of two angles.

23. From theorem 7.14 prove that for any $-\pi/2 < p < \pi/2; -\pi/2 < q < \pi/2$ it is the case that

$$\sqrt{\cos p \cos q} \leq \cos \frac{p+q}{2}. \tag{7.110}$$

24. What are the necessary and sufficient conditions for p and q that will produce the equal sign in inequality (7.79)?
25. To obtain a formula for the cosine of a sum of three numbers $\cos(\theta + \beta + \psi)$, we use associativity of addition: $\cos(\theta + \beta + \psi) = \cos((\theta + \beta) + \psi)$. Then, we can first apply the identity for the cosine of a sum to obtain an expression that contains a trigonometric functions of $(\theta + \beta)$ and of ψ, and then apply trigonometric identities again to get an expression that uses sines and cosines of θ, β, and ψ. Prove that this derivation produces the same result for different groupings of variables in the original sum, for example $\cos((\theta + \beta) + \psi)$ and $\cos(\theta + (\beta + \psi))$. Do the same for $\sin((\theta + \beta) + \psi)$ and $\sin(\theta + (\beta + \psi))$.
26. Consider the following application of associativity of multiplication:

$$(\cos\theta \cdot \cos\beta) \cdot \cos\psi = \cos\theta \cdot (\cos\beta \cdot \cos\psi). \tag{7.111}$$

Use theorem 7.12 to express products in each side through sums of trigonometric functions. Show that the resulting sums of trigonometric functions continue to be equal.
27. Function $\tan\theta$ is strictly monotonic in the interval $-\pi/2 < \theta < \pi/2$ and has a curious property: for any nonzero rational θ the value of $\tan\theta$ is irrational (see p. 160). Prove that for any a, b, where $a < b$, there is no such function $f(x)$ that is strictly monotonic in the interval $a < x < b$ and that does the opposite trick: for any irrational x the value of $f(x)$ is rational. Does such a function exist if we remove the requirement that it is strictly monotonic?
28. Prove that theorem 7.16 cannot be generalized to all real values of n. (Hint: consider $n = 1/k$.)
29. Prove theorem 7.2 from theorem 7.12.

8

Conic Sections

Conic sections, or conics, are three special kinds of plane curves: the ellipse, the parabola, and the hyperbola. They have been studied since antiquity. Over centuries of research, mathematicians proved many beautiful theorems about conics. It is these theorems that make conics a good subject for studying proofs.

There are four different but equivalent definitions for conics:

1. As an intersection of a cone and a plane.
2. A metric definition.
3. An algebraic definition.
4. A definition that uses a focus and a directrix.

A part of this chapter will be dedicated to proving that all these definitions are equivalent. Let's consider them in turn.

Historically, a major application of conic sections was the solution of the so-called two-body problem. Isaac Newton proved that the force of gravity between two spherical or point masses makes them move along conic section curves.

Another application is the design of reflectors for telescopes and telecommunication antennas that we will consider in section 8.10.

Probably, the most common but least recognized application deals with optimization problems, when we need to find a minimum of a function of multiple variables. Many computer optimization algorithms start from an initial guess and then improve on that guess by "walking" in the direction, where the function decreases, until they reach an approximate minimum, where the function is "flat."

Consider a function of two variables in the vicinity of a minimum and imagine a contour elevation plot of this function, like that for a mountainous terrain on a map. Turns out, for a wide variety of functions the elevation contours in the vicinity of the points of interest are described by conics. Figure 8.1 illustrates this fact. It shows contour plots for the function $z = 0.1\sin^2(x) + 1 - \cos(y)$. It has minima $z = 0$ at $x = 0, y = 0$

DOI: 10.1201/9781003456766-8

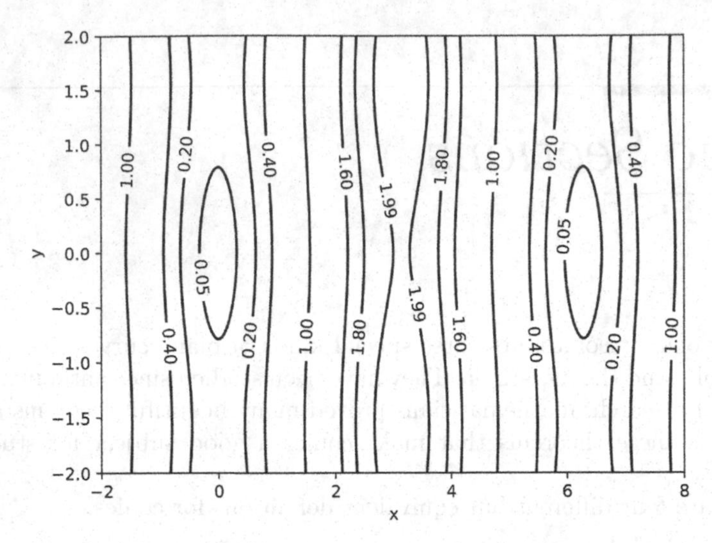

FIGURE 8.1
Plot contours as approximated by conics.

and at $x = 2\pi, y = 0$. Near these minima, elevation contours are approximated by ellipses. Near the point $x = \pi, y = 0$ the constant elevation contours of the function can be approximated by a hyperbola.

This example illustrates a typical behavior: near minima, maxima or other "flat" points, the function's elevation contours are approximated by conics. By using the properties of conics, we can speed up optimization algorithms.

8.1 Cone and Plane Definitions

Conic sections are the curves that are formed by an intersection of a plane with a double cone (hence the name). Depending on the angle between the plane and the axis of the cone, we get an ellipse, parabola, or hyperbola:

1. Figure 8.2A shows a plane crossing one half of the cone and forming an ellipse.
2. If the plane is tilted in such a way that it slices both halves of the cone (figure 8.2C), we get a hyperbola. In contrast to the case of an ellipse, a hyperbola is not a closed curve.

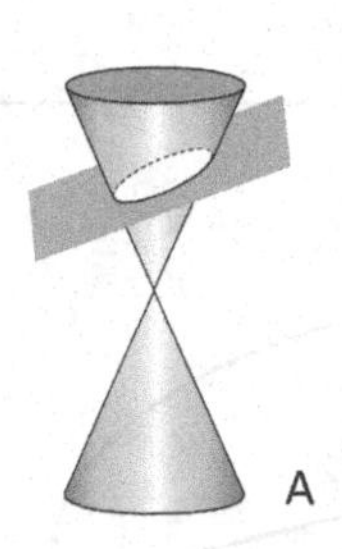

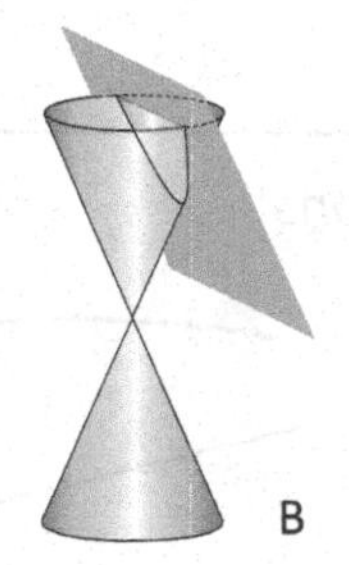

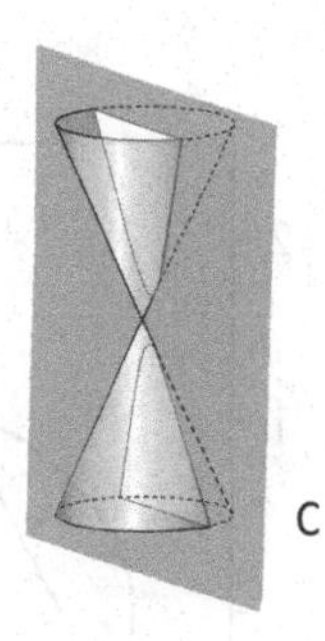

FIGURE 8.2
Conic sections

3. Separating these two cases is a special setting when the plane is parallel to one of the straight lines that lie on the surface of the cone and goes through its vertex. This is a boundary case that shares some features of both the ellipse and the hyperbola. Like in the case of the ellipse, such a plane crosses only one half of the cone, and like in the case of the hyperbola, the cross-section between the plane and the cone is not a closed curve (figure 8.2B). This is the case of parabola.

8.2 Metric Definitions

The title of this sections implies that conic sections can be defined through distances. These definitions are as follows.

1. An ellipse is defined by two points (each called a *focus*) and a distance D that is greater than or equal to the distance between these two points. Then the ellipse is a family of points on the plane, such that the sum of distances from each of these points to the two foci is equal to D. In figure 8.3, the distance $D = d_1 + d_2$ is constant for all points on the ellipse.
2. A hyperbola is defined similarly to the ellipse, but now the *difference* of distances to the two foci is constant. In notations in figure 8.4 we have $d_1 - d_2 = D =$ const for all points on the right branch of the curve. For points on the left branch, $d_1 - d_2 = -D$.
3. For a parabola, the usual metric definition deals with the distance to a point and to a straight line. This definition is given in the next section, along with similar definitions for the ellipse and the hyperbola.

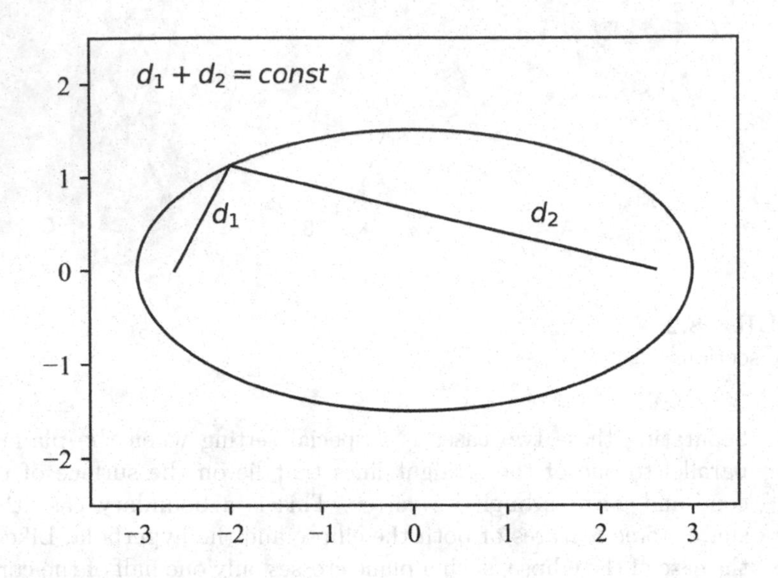

FIGURE 8.3
Ellipse: metric definition

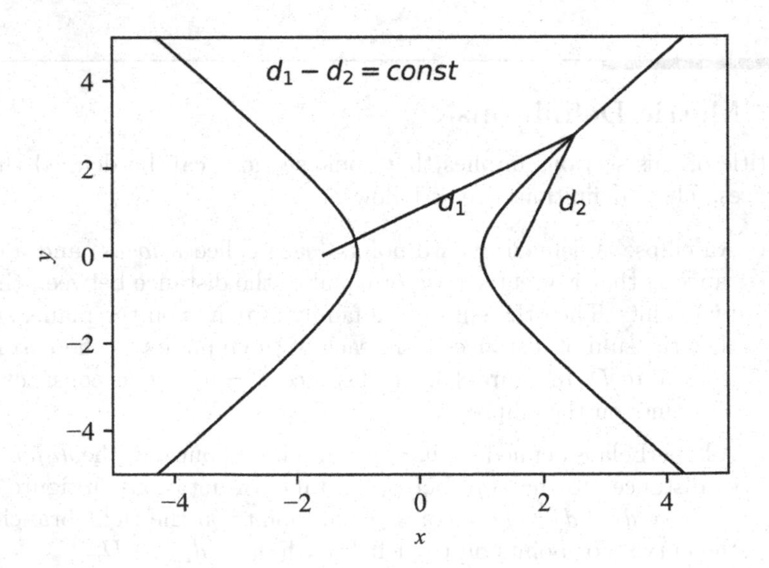

FIGURE 8.4
Hyperbola: metric definition

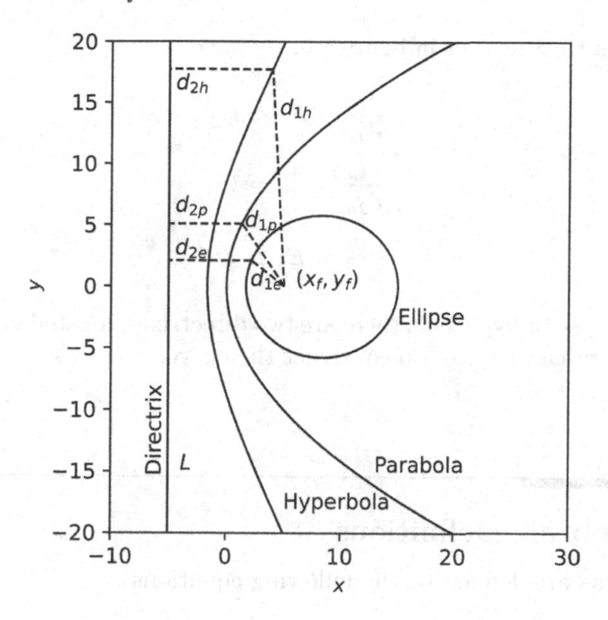

FIGURE 8.5
Conics: the focus and directrix definition

8.3 Focus and Directrix Definitions

On a plane we define a point (x_f, y_f), called the *focus*, and a straight line L, called the *directrix*, in such a way that the focus does not lie on the directrix (figure 8.5). We also define a positive constant E, called the *eccentricity*. Consider a point on the plane (x, y). We denote the distance from point (x, y) to the focus as d_1 and the distance to the directrix as d_2. Let's select all points on the plane that satisfy the following equation:

$$\frac{d_1}{d_2} = E. \tag{8.1}$$

Thus, we consider all points (x, y) for which the ratio of the distance to the focus to the distance to the directrix is constant. Depending on the value of E, points (x, y) will form the following conic sections:

1. For $0 < E < 1$ we get an ellipse.
2. For $E = 1$ we get a parabola.
3. For $E > 1$ we get a hyperbola.

Along the three curves in figure 8.5, we have

$$\frac{d_{1e}}{d_{2e}} = E_e < 1,$$
$$\frac{d_{1p}}{d_{2p}} = E_p = 1, \tag{8.2}$$
$$\frac{d_{1h}}{d_{2h}} = E_h > 1.$$

For an ellipse or hyperbola there are two directrixes, located symmetrically. Either of them can be used to construct the curve.

8.4 Algebraic Definitions

Conic sections are defined by the following equations[1]

1. Ellipse:

$$\frac{x^2}{a^2} + \frac{y^2}{b^2} = 1. \tag{8.3}$$

   (Note that if $a = b$, this equation reduces to $x^2 + y^2 = a^2$, which is a circle of radius a.)

2. Hyperbola:

$$\frac{x^2}{a^2} - \frac{y^2}{b^2} = 1. \tag{8.4}$$

3. Parabola:

$$y^2 = 2px. \tag{8.5}$$

In these equations, a, b, and p are parameters.

> The terms ellipsis, parable, and hyperbole come from linguistics and rhetoric. The ellipsis is the omission of words from speech or text, the parable is a short story to illustrate a moral principle, and the hyperbole is an exaggeration. Note the progression here: from an omission to correspondence (between a story and the principle being illustrated) to superfluity.

[1]These are canonical equations for conics. In fact, any equation in the form

$$ax^2 + bxy + cy^2 + dx + fy + g = 0 \tag{8.6}$$

where $b^2 - 4ac \neq 0$, defines a conic section curve on a plane. Such a curve is still an ellipse, parabola, or hyperbola, but may be shifted and rotated compared to the ones described by the canonical equation.

Is there a similar progression for conic sections called the ellipse, parabola, and hyperbola? As we see from figure 8.2, the angle between the plane and the cone axis varies from the case of ellipse, through parabola and to hyperbola. In another observation, the eccentricity is less than 1 for the ellipse, equals to 1 for parabola, and is greater than 1 for hyperbola.

8.5 Equivalency of Definitions 1 and 2

Theorem 8.1 *The definitions of conics via a cone and a plane are equivalent to their metric definitions.*

The original proof of this theorem was given in late antiquity and was quite cumbersome, but in 1822 the French mathematician Germinal Pierre Dandelin found an elegant visual proof. Here, we provide the proof only for the ellipse. Proofs for the hyperbola and parabola are the subject of exercises 1 and 2 for this chapter.

Proof

First, we make an observation. Consider a sphere and two lines that are coming from the same point and are tangential to the sphere (figure 8.6). We note that line segments AB and AC have the same length: $|AB| = |AC|$.

The main body of the proof is as follows. We consider a cone and a plane that intersects the cone. We place two spheres (called Dandelin spheres) in such a way, that each of them is tangent to the plane and the cone, and that

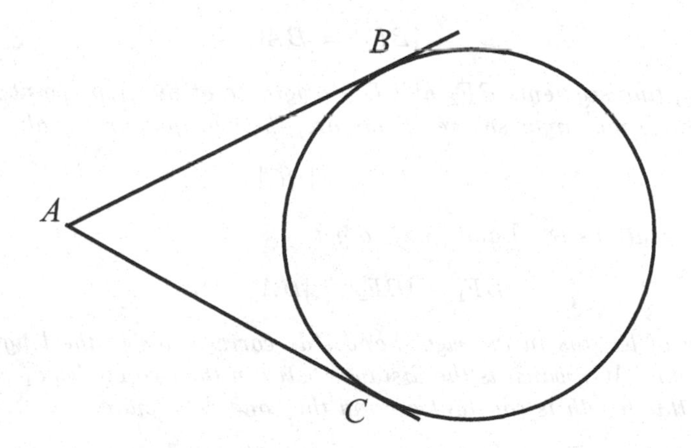

FIGURE 8.6
A sphere and two tangential lines

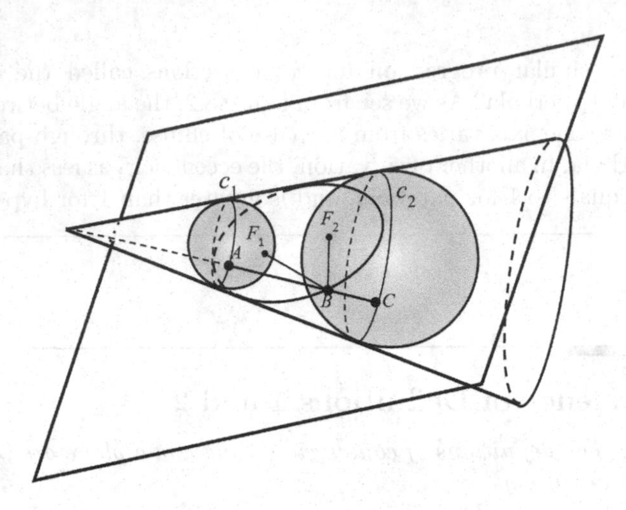

FIGURE 8.7
The Dandelin spheres: ellipse

they are located on both sides of the plane (figure 8.7, where the smaller sphere is on the top side of the plane, and the larger sphere is on the bottom side). These spheres touch the cone at circles c_1 and c_2. We also denote the points where the plane touches the spheres as F_1 and F_2.

The plane and the cone intersect at some curve that is shown in the figure. Consider a point B on that curve. We draw a line on the cone from the vertex to point B and extend it to circle c_2. This line intersects circles c_1 and c_2 at points A and C.

Line segments BF_1 and BA originate at the same point B and are tangential to the smaller sphere. Therefore, their lengths are equal:

$$|BF_1| = |BA|. \tag{8.7}$$

Similarly, line segments BF_2 and BC originate at the same point B and are tangential to the larger sphere. Therefore, their lengths are equal:

$$|BF_2| = |BC|. \tag{8.8}$$

We add equations (8.7) and (8.8) to get

$$|BF_1| + |BF_2| = |BA| + |BC|. \tag{8.9}$$

The sum of lengths in the right-hand side corresponds to the length $|AC|$ of line segment AC, which is the distance between the two circles c_1 and c_2. We see that this length is constant around the cone. Therefore,

$$|BF_1| + |BF_2| = const, \tag{8.10}$$

which is the metric definition of the ellipse.

8.6 Equivalency of Definitions 2 and 4

In this section we prove that algebraic definitions of conics are equivalent to their metric definitions.

Theorem 8.2 *The algebraic and metric definitions of conics are equivalent.*

Proof

We consider the ellipse and hyperbola here. The metric definition for parabola is a bit different, and we consider it along with the directrix and focus definitions for conics.

1. *The algebraic definition states that points on an ellipse satisfy the following equation*

$$\frac{x^2}{a^2} + \frac{y^2}{b^2} = 1. \tag{8.11}$$

The foci of this ellipse are located at

$$\begin{aligned} x_f &= \pm\sqrt{a^2 - b^2}, \\ y_f &= 0. \end{aligned} \tag{8.12}$$

The metric definition states that for every point on the ellipse, the sum of distances from that point to the foci is constant and equal to 2a (figure 8.3 on p. 188). Mathematically this is expressed as follows (here we denote $c = \sqrt{a^2 - b^2}$ for brevity):

$$d_1 + d_2 = 2a, \tag{8.13}$$

where

$$\begin{aligned} d_1 &= \sqrt{(x+c)^2 + y^2}, \\ d_2 &= \sqrt{(x-c)^2 + y^2}. \end{aligned} \tag{8.14}$$

We substitute expressions for d_1, d_2 in equation (8.13) and move one of the radicals to the right-hand side:

$$\sqrt{(x+c)^2 + y^2} = 2a - \sqrt{(x-c)^2 + y^2}. \tag{8.15}$$

Squaring this equation yields

$$(x+c)^2 + y^2 = 4a^2 + (x-c)^2 + y^2 - 4a\sqrt{(x-c)^2 + y^2}. \tag{8.16}$$

We expand the parentheses and cancel some of the resulting terms

$$2xc = 4a^2 - 2xc - 4a\sqrt{(x-c)^2 + y^2}. \tag{8.17}$$

Next, we divide both sides by four and isolate the radical

$$a^2 - xc = a\sqrt{(x-c)^2 + y^2}. \tag{8.18}$$

We square the equation again to remove the radical:

$$a^4 - 2a^2xc + x^2c^2 = a^2(x-c)^2 + a^2y^2. \tag{8.19}$$

After expanding the parentheses some terms cancel:

$$a^4 - 2a^2xc + x^2c^2 = a^2x^2 - 2a^2xc + a^2c^2 + a^2y^2. \tag{8.20}$$

Then

$$a^2(a^2 - c^2) = x^2(a^2 - c^2) + a^2y^2. \tag{8.21}$$

We recall that $c = \sqrt{a^2 - b^2}$. *This produces*

$$a^2b^2 = x^2b^2 + a^2y^2. \tag{8.22}$$

We divide this equation by a^2b^2 *to get the algebraic definition of the ellipse.*

2. *For hyperbola, we consider the difference of the distances to focal points (see figure 8.4 on p. 188). The proof is similar to that for the ellipse. We start from*

$$d_1 - d_2 = 2a, \tag{8.23}$$

where

$$\begin{aligned} d_1 &= \sqrt{(x+c)^2 + y^2}, \\ d_2 &= \sqrt{(x-c)^2 + y^2}, \end{aligned} \tag{8.24}$$

and where the foci are located at $x_f = \pm c$. *We follow the same line of algebraic transformations as in the case for ellipse to get*

$$x^2(c^2 - a^2) - a^2y^2 = a^2(c^2 - a^2). \tag{8.25}$$

We denote $b^2 = c^2 - a^2$ *and divide the last equation by* a^2b^2 *to get equation (8.4).*

8.7 Equivalency of Definitions 3 and 4

In this section we prove that the algebraic definitions of conics are equivalent to their definitions via the focus and directrix. We start from the case of a parabola, which is a bit less cumbersome.

Theorem 8.3 *The algebraic definition of a parabola is equivalent to its definition that uses the focus and directrix.*

Proof

We define the parabola as

$$y^2 = 2px. \tag{8.26}$$

We also define the location of the focus $x_f = p/2, y_f = 0$ and the directrix at $x_d = -p/2$ (see the parabola curve in figure 8.5). We recall that for a parabola the eccentricity is equal to one. Therefore, for any point (x, y) on the parabola, the distance from that point to the focus should be equal to the distance to the directrix. Let's check if that is the case.

The distance d_{2p} from point (x, y) on the parabola to the directrix (see figure 8.5) is given by

$$d_{2p} = x - x_d = x + \frac{p}{2}. \tag{8.27}$$

The distance d_{1p} from point (x, y) on the parabola to the focus is given by

$$d_{1p} = \sqrt{(x - x_f)^2 + y^2} = \sqrt{\left(x - \frac{p}{2}\right)^2 + y^2}. \tag{8.28}$$

We plug the expression for y^2 from equation (8.26) into the expression for d_{1p} to get

$$d_{1p} = \sqrt{\left(x - \frac{p}{2}\right)^2 + 2px}. \tag{8.29}$$

After expanding the parentheses and collecting the terms we get

$$d_{1p} = \sqrt{x^2 + px + \frac{p^2}{4}}. \tag{8.30}$$

Under the radical we recognize a complete square:

$$d_{1p} = \sqrt{\left(x + \frac{p}{2}\right)^2}. \tag{8.31}$$

A comparison of equations (8.27) and (8.31) shows that

$$d_{1p} = d_{2p}. \tag{8.32}$$

Next, we proceed to the case of ellipse.

Theorem 8.4 *The algebraic definition of the ellipse is equivalent to its definition via the focus and directrix, except when the ellipse degenerates into a circle.*

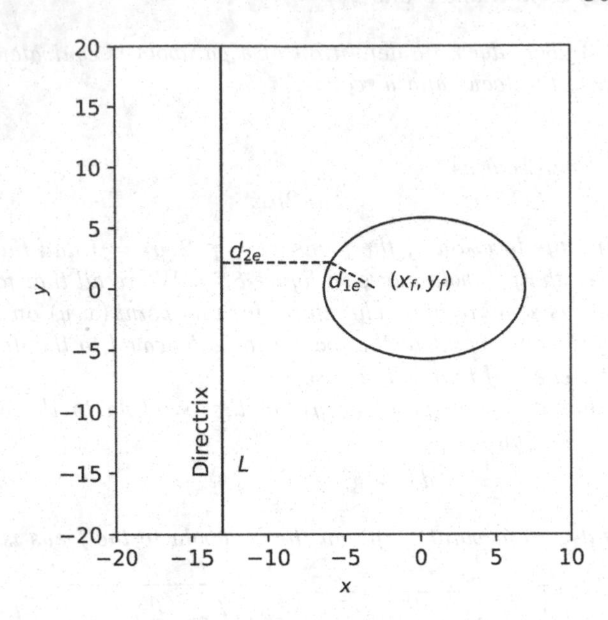

FIGURE 8.8
Ellipse: directrix and focus definition

Proof
Algebraically, the ellipse is defined by the following equation:

$$\frac{x^2}{a^2} + \frac{y^2}{b^2} = 1. \tag{8.33}$$

Here, we assume that $a > b$. The case $a < b$ is analogous if we swap x and y axes. (The case $a = b$ is not considered because a circle is excluded from the formulation of this theorem.)

For this ellipse, the foci are located at $x_f = \pm c, y_f = 0$, where $c = \sqrt{a^2 - b^2}$. The eccentricity is equal to $E = \sqrt{a^2 - b^2}/a$ (see figure 8.8).

We select the focus $x_f = -c$; the case for the other focus is analogous. For this focus, the directrix is located at $x_d = -a^2/c$. We select a point (x, y) on the ellipse and compute the distances to the focus and the directrix. The distance d_{1e} to the focus is given by

$$d_{1e}^2 = (x + c)^2 + y^2. \tag{8.34}$$

A solution of equation (8.33) for y^2 is as follows:

$$y^2 = b^2 - \frac{b^2}{a^2}x^2 \tag{8.35}$$

A substitution of this value in the right-hand side of equation (8.34) yields:

$$d_{1e}^2 = (x + c)^2 + b^2 - \frac{b^2}{a^2}x^2. \tag{8.36}$$

We expand $(x+c)^2$ and use the definition of c:

$$d_{1e}^2 = x^2 + 2x\sqrt{a^2-b^2} + a^2 - b^2 + b^2 - \frac{b^2}{a^2}x^2 \tag{8.37}$$

or

$$d_{1e}^2 = x^2\frac{a^2-b^2}{a^2} + 2x\sqrt{a^2-b^2} + a^2. \tag{8.38}$$

This happens to be a complete square:

$$d_{1e}^2 = \left(x\frac{\sqrt{a^2-b^2}}{a} + a\right)^2. \tag{8.39}$$

Therefore,

$$d_{1e} = x\frac{\sqrt{a^2-b^2}}{a} + a. \tag{8.40}$$

Now, we proceed to the computation of the distance to the directrix. It is given by

$$d_{2e} = x + \frac{a^2}{c}. \tag{8.41}$$

Here, we use the definition of c to get

$$d_{2e} = x + \frac{a^2}{\sqrt{a^2-b^2}}. \tag{8.42}$$

A comparison of equations (8.40) and (8.42) shows that

$$d_{1e} = d_{2e}\frac{c}{a}. \tag{8.43}$$

Therefore,

$$\frac{d_{1e}}{d_{2e}} = \frac{c}{a} = E. \tag{8.44}$$

Note that $E < 1$ because $c = \sqrt{a^2-b^2} < a$.

Finally, we consider the case of hyperbola. It turns out to be largely analogous to the case of ellipse, with a few signs flipped.

Theorem 8.5 *Algebraic definition of hyperbola is equivalent to its definition via the focus and directrix.*

Proof

Algebraically, the hyperbola is defined by the following equation:

$$\frac{x^2}{a^2} - \frac{y^2}{b^2} = 1. \tag{8.45}$$

The foci are located at $x_f = \pm c, y_f = 0$, where $c = \sqrt{a^2+b^2}$. The eccentricity is equal to $E = \sqrt{a^2+b^2}/a$. Note that now $E > 1$. We select the focus $x_f = c$;

the case for the other focus is analogous. For this focus, the directrix is located at $x_d = a^2/c$. *We select a point* (x, y) *on the hyperbola and compute the distances to the focus and the directrix. The distance to the focus is given by*

$$d_{1h}^2 = (x - c)^2 + y^2. \tag{8.46}$$

Similarly to the ellipse case, we solve for y^2 *in equation (8.33):*

$$y^2 = -b^2 + \frac{b^2}{a^2}x^2. \tag{8.47}$$

Note that signs in the right-hand side are flipped compared to the case of ellipse. The rest of the derivation is completely analogous. We get

$$\frac{d_{1h}}{d_{2h}} = \frac{c}{a} = E, \tag{8.48}$$

where $E > 1$ *now.*

8.8 Conics in Polar Coordinates

Algebraic equations (8.3), (8.4), and (8.5) for the three conics look different, but we also see that the focus and directrix definitions for these curves differ only by the value of the eccentricity. Can we derive a formula that would uniformly apply to all conics? The answer is given by the following theorem.

Theorem 8.6 *In polar coordinates, equations for the ellipse, parabola, and one branch of the hyperbola are given by*

$$r = \frac{p}{1 - E\cos\theta}, \tag{8.49}$$

where E *is the eccentricity and* p *is defined as follows:*

1. *For the ellipse and hyperbola,* $p = b^2/a$, *with* a *and* b *defined by equations (8.3) and (8.4).*
2. *For the parabola,* p *is defined by equation (8.5).*

As a reminder, $E < 1$ for an ellipse, $E = 1$ for a parabola, and $E > 1$ for a hyperbola. In all cases, $p = EH$, where H is the distance between the focus and the directrix.

Proof

We limit the proof to the ellipse, parabola, and one branch of hyperbola. We place the origin of the coordinate system at a focus (figure 8.9). This placement of the origin is shifted compared to those in section 8.7. Therefore, the numerical values of the positions of the directrix and focus will be different from those above. As a reminder, the old locations of the focus and directrix were (see section 8.7):

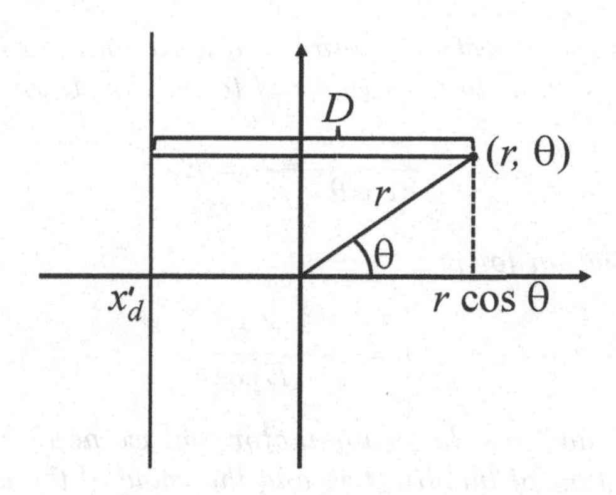

FIGURE 8.9
Focus and directrix in polar coordinates

1. *For parabola,* $x_f = p/2, x_d = -p/2$.
2. *For ellipse,* $x_f = -c, x_d = -a^2/c$.
3. *For hyperbola,* $x_f = c, x_d = a^2/c$.

The ellipse and the hyperbola have two directrixes (see p. 190); choosing a particular directrix for the proof of this theorem is a matter of convenience. In the new axes, the focus is located at $(0, 0)$ *for all three curves. The directrix is located at:*

1. *For the parabola,* $x'_d = -p$.
2. *For the ellipse,* $x'_d = -a^2/c + c$. *This formula is simplified as*

$$\begin{aligned} x'_d &= -\frac{a^2}{c} + c \\ &= -\frac{a^2 - c^2}{c} \\ &= -\frac{a^2 - (a^2 - b^2)}{c} \\ &= -\frac{b^2}{c}. \end{aligned} \tag{8.50}$$

3. *For hyperbola,* $x'_d = a^2/c - c$, *and a similar simplification also produces* $-b^2/c$.

Consider the point (r, θ) *on the curve. The distances from this point to the focus is simply given by* r *(here the polar coordinates work to our advantage). The distance from point* (r, θ) *to the directrix is given by*

$$D = r\cos\theta - x'_d. \tag{8.51}$$

We use the focus and directrix definition to write an equation for the ratio of distances from this point to the focus and to the directrix:

$$\frac{r}{r\cos\theta - x'_d} = E. \tag{8.52}$$

We solve this equation for r:

$$r = \frac{-x'_d E}{1 - E\cos\theta}. \tag{8.53}$$

This equation already has the denominator that we need. For the numerator, we recall the position of the directrix and the value of the eccentricity:

1. *For parabola, $x'_d = -p$ and $E = 1$.*
2. *For ellipse and hyperbola, $x'_d = -b^2/c$ and $E = c/a$.*

In all cases we get the value of p as defined by the theorem.

8.9 Ray Reflections by Conics

Here, we consider a geometric property that has a counterpart in the real world. For a ray of light that is reflected by a mirror, the angle of incidence is equal to the angle of reflection. If the mirror is curved, then each angle is defined as the angle between the ray and the plane that is tangent to the mirror at the reflection point.

Imagine a mirror whose cross section has the shape of a conic section. Rays of light that are reflected from such a mirror have interesting properties:

Theorem 8.7 *A ray coming from one focus of the ellipse and reflected by the ellipse goes through the second focus.*

This theorem is illustrated in figure 8.10. Geometrically, this means that two lines from the foci to a point on the ellipse form equal angles with the tangent line: $\alpha = \beta$.

Before we prove this theorem, we prove

Lemma 8.1 *Figure 8.11 shows two line segments, AX and XB. Consider the combined length $|AX| + |XB|$ of these two segments as a function of the position of point X on line l. Then the combined length $|AX| + |XB|$ achieves its minimum when $\alpha = \beta$.*

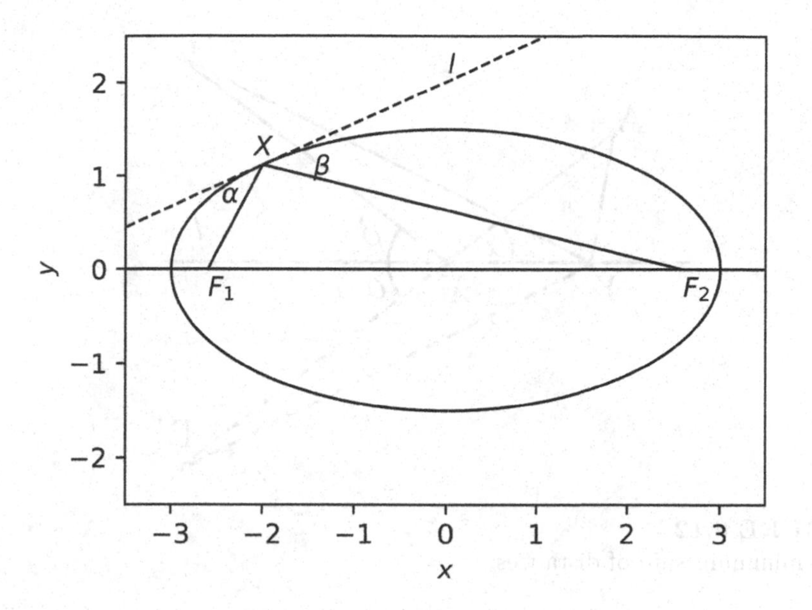

FIGURE 8.10
Ellipse ray reflection property

Proof
Consider figure 8.12. We construct a new point B' that is the reflection of point B with respect to line l. Consider two points, X and Y. We have

$$\begin{aligned} |AX| + |XB| &= |AX| + |XB'|, \\ |AY| + |YB| &= |AY| + |YB'|. \end{aligned} \tag{8.54}$$

We also note that if points A, X, and B' lie of the same line, then

$$|AX| + |XB'| \leq |AY| + |YB'| \tag{8.55}$$

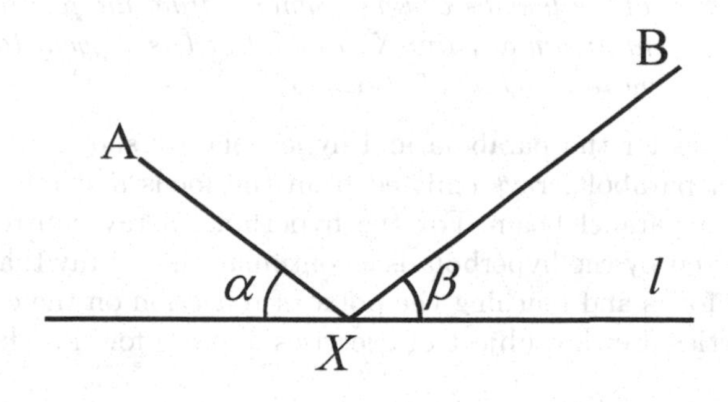

FIGURE 8.11
Ray reflection lemma

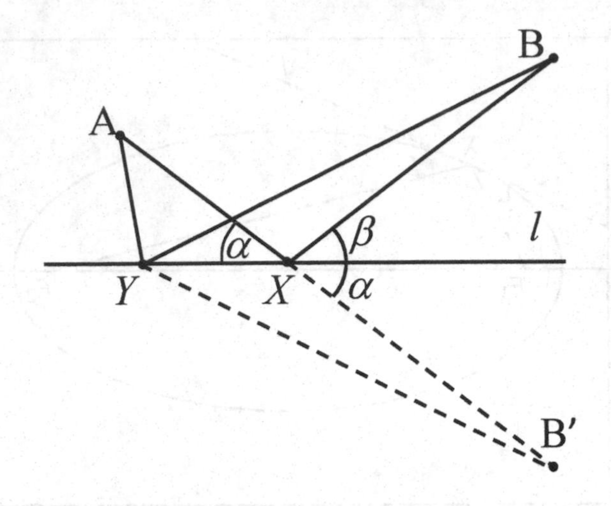

FIGURE 8.12
The minimum sum of distances

Then

$$|AX| + |XB| \leq |AY| + |YB|. \tag{8.56}$$

In that case it is true that the angle between AX and line l is equal to the angle between XB and line l.

Now, we can proceed to proving the ray property for the ellipse.

Proof

We again refer to figure 8.10. We know that the sum of lengths $|F_1X| + |F_2X|$ is constant for all points on the ellipse. Therefore, for all points outside the ellipse, the sum of lengths from that point to the foci will be greater than $|F_1X| + |F_2X|$.

Now consider a line l that is tangential to the ellipse. All points on that line lie outside of the ellipse, except a single point that lies on the ellipse. Therefore, the sum of the lengths of line segments from the foci to a point on line l will have the minimum at point X, where line l is tangent to the ellipse. We invoke the lemma to conclude that $\alpha = \beta$.

Ray properties for the parabola and hyperbola are shown in figures 8.13 and 8.14. For a parabola, rays emitted from the focus and reflected by the parabola form a parallel beam. For the hyperbola, a ray emitted from one focus and reflected by the hyperbola is a continuation of a ray that is emitted from the other focus and reaching the point of reflection on the curve. Proofs of these properties are the subject of exercises 4 and 5 for this chapter.

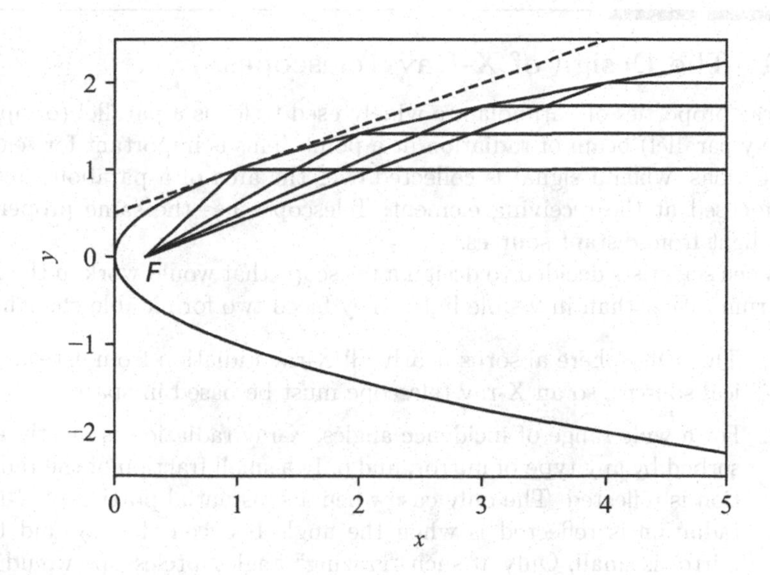

FIGURE 8.13
Ray properties for parabola

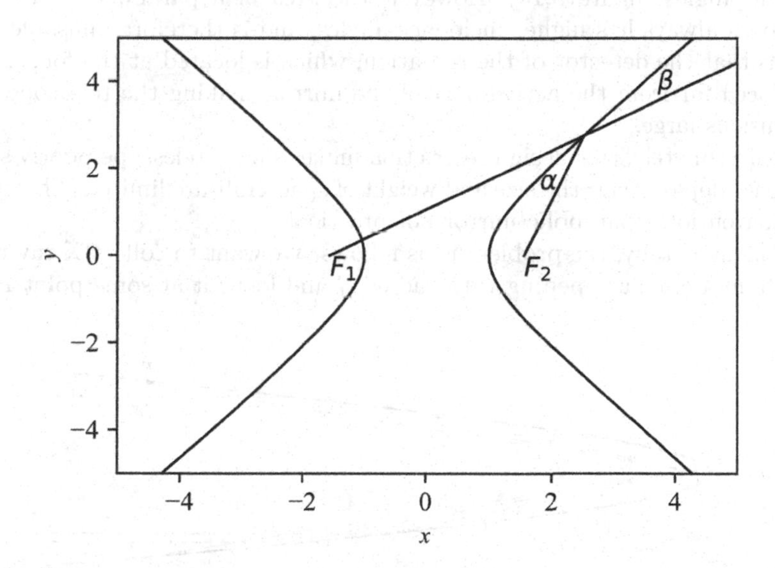

FIGURE 8.14
Ray properties for hyperbola

8.10 The Design of X-Ray Telescopes

The ray properties of parabola are widely used to focus a parallel (or approximately parallel) beam of radiation at a point. This is important for telecommunications, when a signal is collected over the area of a parabolic antenna and focused at the receiving element. Telescopes use the same property to focus light from distant sources.

When scientists decided to design a telescope that would work in the X-ray spectrum rather than in visible light, they faced two formidable challenges:

1. The atmosphere absorbs nearly all X-ray radiation from astronomical sources, so an X-ray telescope must be based in space.
2. For a wide range of incidence angles, X-ray radiation is mostly absorbed by any type of mirror, and only a small fraction of the radiation is reflected. The only case when a substantial portion of X-ray radiation is reflected is when the angle between the ray and the mirror is small. Only at such "grazing" angles a telescope would be able to capture enough power of the incoming X-rays to form an image.

Let us consider the second challenge first. Of course, if a parabola is extended sufficiently far, a part of the mirror does collect the radiation with grazing angles (figure 8.15). However, the area of a parabolic mirror near the focus always has higher incidence angles, and is therefore unusable. This means that the detector of the radiation, which is located at the focus, must be placed far from the active part of the mirror, making the telescope axial dimensions large.

Unfortunately, this design restriction makes such a telescope poorly suited for space deployment: the size and weight of spacecraft are limited. This makes the conventional parabolic mirror not practical.

Fundamentally, the problem is as follows. We want to collect X-ray radiation from a circular opening with radius R and focus it at some point F. We

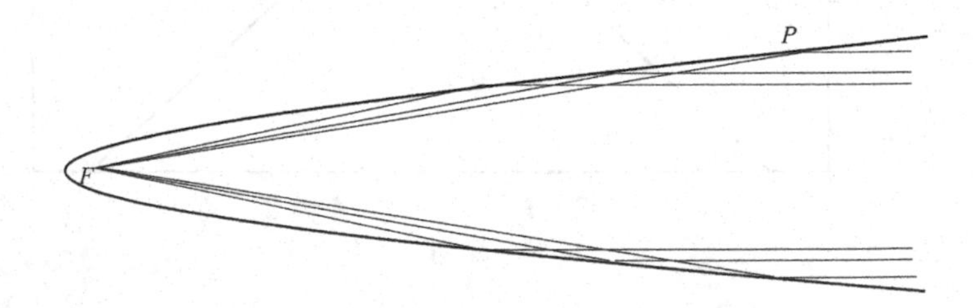

FIGURE 8.15
Parabolic reflector for an X-ray telescope

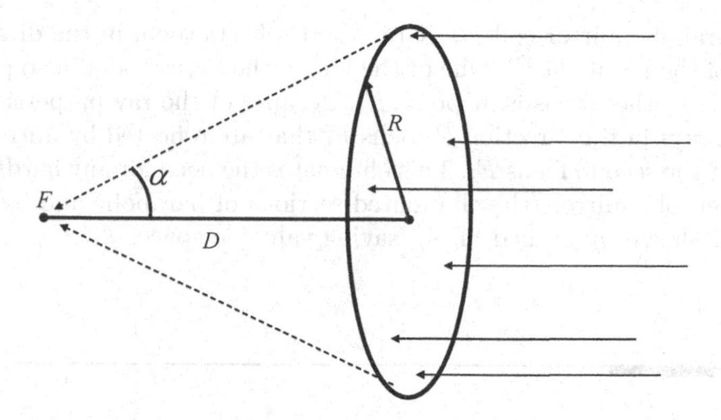

FIGURE 8.16
Design constraints for an X-ray telescope

also want to keep the axial dimension D of the device to a minimum. Figure 8.16 shows the geometry of this problem without specifying a way to deflect incoming rays to point F. From this figure we see that some of the rays at the periphery of the opening must be somehow deflected by angle α that is given by $\alpha = \tan^{-1} R/D$. Note that we do not specify the exact way a ray must be deflected; the constraint on α and D stems from the general requirement to focus X-ray radiation from an opening at some point F.

We see that for smaller D, angle α can be large. For a telescope, this cannot be achieved by a simple reflection because of the small reflection coefficient for such an angle. The solution is to realize that the net change in the direction of a ray does not have to be a result of a single reflection. We can have two or more consecutive reflections, each deflecting the ray by a smaller angle.

Today, X-ray telescopes use the so-called Wolter design that makes a good use of grazing angles while keeping the dimensions of the telescope down. A cross-section of a Wolter telescope is shown in figure 8.17. It combines two mirrors: a section of a parabolic mirror P and a section of a hyperbolic mirror

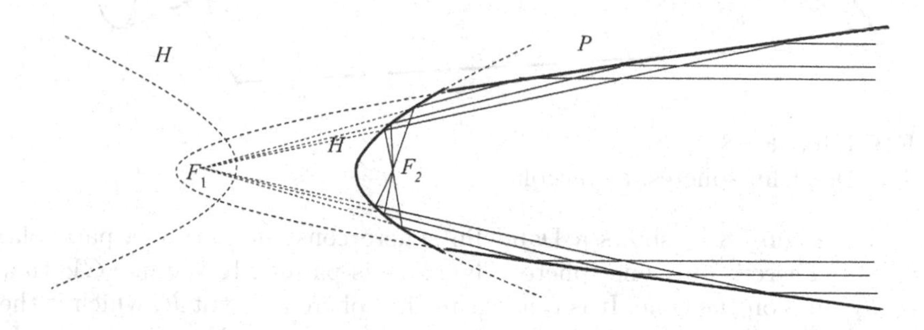

FIGURE 8.17
Telescope design for grazing angles

H. The parabolic mirror collects X-rays and reflects them in the direction of the focus of the parabola F_1. One of the foci of the hyperbola is also placed at point F_1. The other focus is at point F_2. Because of the ray properties of the hyperbola, rays in the direction of focus F_1 that are reflected by mirror H are collected at the second focus F_2. This eliminates the need for any hardware left of the hyperbolic mirror (the eliminated sections of parabolic and hyperbolic mirrors are shown by dashed lines), saving valuable space.

Exercises

1. For hyperbola, Dandelin spheres are placed in the different halves of the cone (figure 8.18). Point A is located on the conic curve—that is, on the intersection of the cone and the plane. The plane is tangential to the spheres at points F_1 and F_2, and (unlike the case of the ellipse) the spheres are located on the same side of the plane. Points B and C are on the circles where the spheres are tangent to the cone.
   Using this construction, prove that the metric definition for hyperbola holds:
   $$|AF_2| - |AF_1| = \text{const}, \tag{8.57}$$

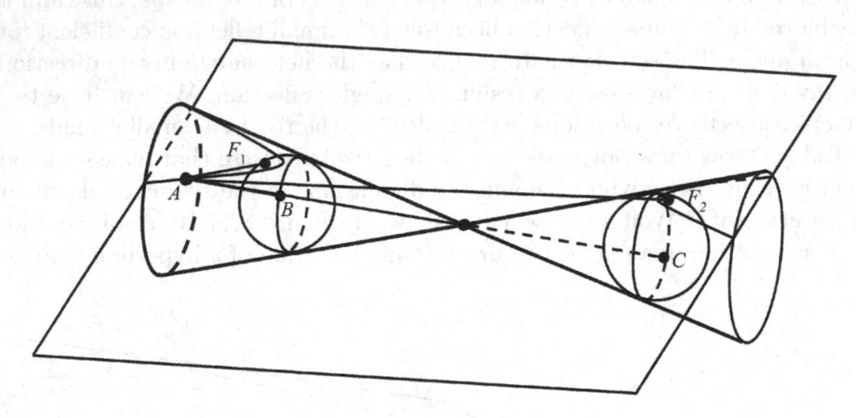

FIGURE 8.18
The Dandelin spheres: hyperbola

2. Figure 8.19 shows a Dandelin sphere construction for a parabola. There is only one sphere. The plane is parallel to the line GV that lies on the cone. It is tangent to the sphere at point F, which is the focus for the parabola. The sphere is tangent to the cone at circle L. The plane that contains circle L and the plane that crosses the cone intersect at line d, which is the directrix for the parabola. We select

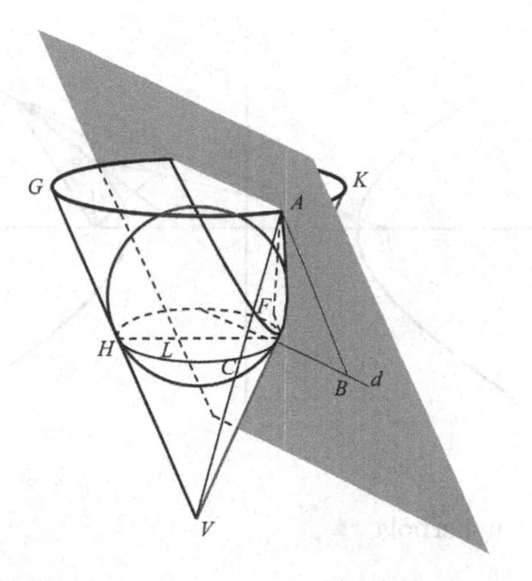

FIGURE 8.19
The Dandelin sphere: parabola

a point A on the conic curve and consider circle K that contains that point A.

Prove that this conic section satisfies the focus and directrix definition for the parabola—that is, $AF = AB$. (Hint: compare lengths of AF, AC, GH, and AB.)

3. A proof of equivalence of algebraic and metric definitions for hyperbola starting on p. 194 misses a chain of algebraic calculations. Recreate the full derivation and complete this proof.
4. Figure 8.20 presents a geometric construction for proving the ray property of parabola. Consider a point A on the parabola and draw

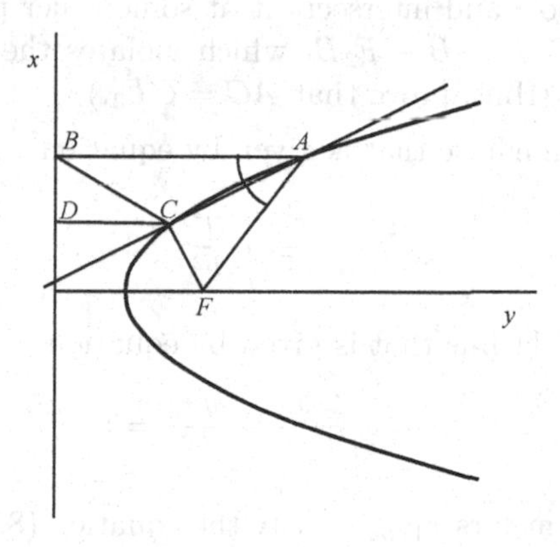

FIGURE 8.20
Ray properties for parabola

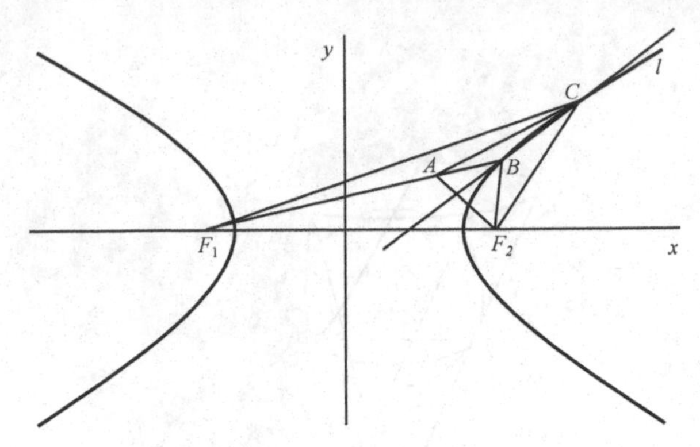

FIGURE 8.21
Ray properties for hyperbola

line AF to the focus and line AB that is parallel to the axis y and intersects the directrix at point B.

The proof is by contradiction. Assume that the line bisecting the angle BAF is not tangent to the parabola, but intersects it at some point C. If that point lies on the parabola, we must have $CD = CF$. Prove that this is false, pointing to a contradiction.

5. Figure 8.21 is used to prove the ray property of hyperbola. Consider foci F_1 and F_2 and a point on the hyperbola B. Draw line l bisecting the angle F_1BF_2. Select point A on the line F_1B, such that $AB = BF_2$. Because of the metric property of the hyperbola, the distance $F_1A = F_1B - BF_2$ must not be dependent on the position of point B on the curve.

   The proof is by contradiction. Assume that line l is not tangent to the hyperbola and intersects it at some other point C. Prove that $F_1C - F_2C \neq F_1B - F_2B$, which violates the metric property of hyperbola. (Hint: prove that $AC = CF_2$.)

6. Consider an ellipse that is given by equation

$$\frac{x^2}{a^2} + \frac{y^2}{b^2} = 1, \tag{8.58}$$

   and a straight line that is given by equation

$$\frac{xx_1}{a^2} + \frac{yy_1}{b^2} = 1, \tag{8.59}$$

   where parameters x_1, y_1 satisfy the equation (8.58) for the ellipse—that is,

$$\frac{x_1^2}{a^2} + \frac{y_1^2}{b^2} = 1. \tag{8.60}$$

Prove that straight line (8.59) is tangential to the ellipse at the point (x_1, y_1). (Hint: add equations (8.58) and (8.60) and then subtract twice equation (8.59) from the result.)

7. We say that variable $\tilde{x}$ is equal to a scaled variable x if $\tilde{x} = \alpha x$, where α is a constant. Consider equation for a circle:

$$x^2 + y^2 = R^2. \tag{8.61}$$

Prove that a scaled equation

$$\tilde{x}^2 + y^2 = R^2 \tag{8.62}$$

defines an ellipse.

8. Prove that a cross-section of a cylinder and a plane is an ellipse. Use two methods:
   (a) Consider a projection of that cross-section on a plane that is perpendicular to the axis of the cylinder, then use the result of problem 7.
   (b) Use Dandelin spheres that are inscribed in the cylinder.
9. Use the algebraic definitions for the ellipse and hyperbola (equations (8.3) and (8.4)). Prove that if $a = b$, then
   (a) An ellipse degenerates into a circle.
   (b) Asymptotes of the hyperbola form a right angle.
10. Consider figure 8.22. The axis of a parabola intersects the directrix at point A. A chord originates at point A and intersects the parabola at points B and C. Prove that angles α and β are equal.

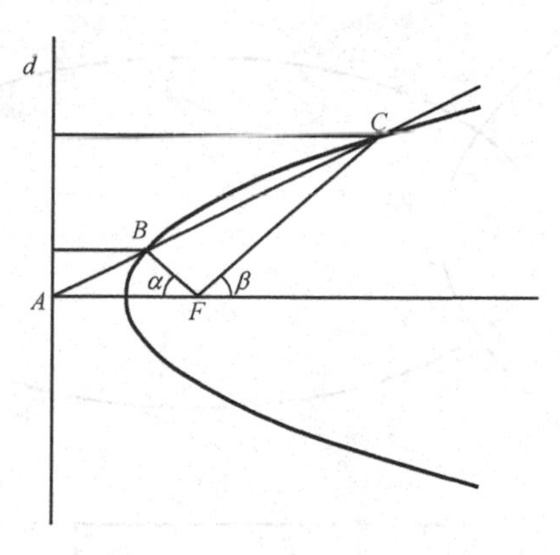

FIGURE 8.22
A conic chord intersects the axis at the directrix

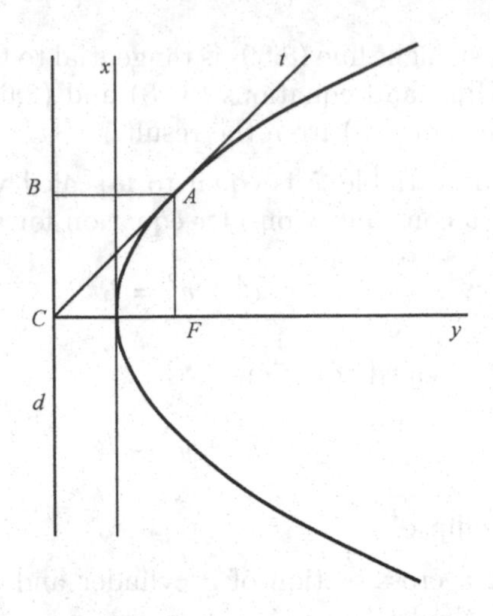

FIGURE 8.23
A tangent to a parabola and the directrix

11. Consider figure 8.23. The axis of a parabola intersects the directrix d at point C. Line t originates at point C and is tangent to the parabola at point A. Line AB is perpendicular to the directrix d. Prove that $CBAF$ is a square.

12. Figure 8.24 shows confocal ellipse and hyperbola—that is, a case when these two curves share their foci. Use the ray properties for the ellipse and hyperbola to prove that they intersect at the right angles.

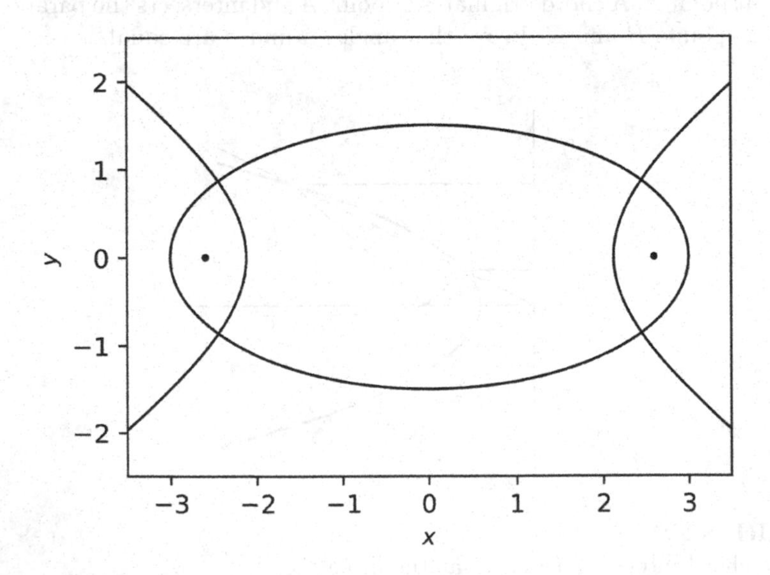

FIGURE 8.24
Confocal conic sections

9

Sequences and Sums

In this chapter, we consider sequences of numbers, like $1, 2, 3, ..., 100$ or $\frac{1}{2}, \frac{1}{3}, \frac{1}{4}, ..., \frac{1}{17}$. We will use a subscript to define individual terms of a sequence: S_n is the n-th element of sequence $\{S\}$. From this explanation we see that a sequence can be viewed as a function with the domain that contains some or all consecutive natural numbers, $f(n)$. Then $f(1)$ will be the first term of a sequence, $f(2)$ the second term, and so on.

For some of the sequences we will also compute sums of all their elements. We start from finite sequences, which means that a sequence has a finite number of terms. After that, we will move to the rich field of infinite sequences and sums.

> A stream of numerical data can be viewed as a number sequence. This defines many applications of sequences in computer science, statistics and other fields. Many other applications are based on the notion of convergence (explained in section 9.4), when the numbers in a sequence become ever closer to some fixed value. The vast majority of practical problems do not allow a closed-form solution. For example, as we mentioned in section 5.8, there is no closed-form general solution for the roots of polynomials of degrees 5 and up. When faced with such intractable problems, scientists and engineers often use sequences that approach (converge to) the exact solution. By computing elements of such a sequences one by one, they can get as close to the exact value as it is necessary for the application in hand. This approach, for example, is used by equation (2.132) to compute the approximate value of a square root.

9.1 Arithmetic Sequence

The arithmetic sequence is defined by its first element S_1 and a recursive relation:

$$S_{n+1} = S_n + D, \tag{9.1}$$

DOI: 10.1201/9781003456766-9

where D is a constant. For example, $1, 2, 3, ..., 127$ is an arithmetic sequence because the difference between successive elements is equal to 1. Another example of arithmetic sequence is $\pi, \pi+0.1, \pi+0.2, ..., \pi+1$. The n-th element of an arithmetic sequence (9.1) is given by

$$S_n = S_1 + (n-1)D. \tag{9.2}$$

Let us now compute the sum of all elements of a finite arithmetic sequence.

Theorem 9.1 *The sum of arithmetic sequence (9.1) is given by*

$$\sum_{n=1}^{N} S_n = NS_1 + D\frac{N(N-1)}{2}. \tag{9.3}$$

Here are two proofs of this theorem.

Proof
Let us compute the sum of the first and the last elements of the sequence:

$$S_1 + S_N = S_1 + S_1 + D(N-1) = 2S_1 + D(N-1). \tag{9.4}$$

The sum of the second element and the one before the last is as follows:

$$S_2 + S_{N-1} = S_1 + D + S_1 + D(N-2) = 2S_1 + D(N-1). \tag{9.5}$$

We see that such pairs always produce the same value $2S_1 + D(N-1)$. *This gives us a way to compute the sum: we just need to determine the number of such pairs. The total number of elements is* N. *If* N *is even, the total number of pairs is* $N/2$. *Then the sum of the sequence is given by*

$$\begin{aligned}\sum_{n=1}^{N} S_n &= (2S_1 + D(N-1))\,\frac{N}{2} \\ &= NS_1 + D\frac{N(N-1)}{2}.\end{aligned} \tag{9.6}$$

A proof of this formula for odd N *is a subject of exercise 20 for this chapter.*

The second proof uses induction.

Proof
The two steps of this proof are:

1. *For* $N = 1$ *formula (9.3) is correct.*
2. *Suppose that formula (9.3) is correct for* N. *Consider the sequence that has one additional element. The value of that element will be* $S_{N+1} = S_1 + ND$. *Then*

$$\begin{aligned}\sum_{n=1}^{N+1} S_n &= NS_1 + D\frac{N(N-1)}{2} + S_1 + DN \\ &= (N+1)S_1 + D\frac{N(N-1)+2N}{2} \\ &= (N+1)S_1 + D\frac{(N+1)N}{2}.\end{aligned} \tag{9.7}$$

This proves the theorem.

9.2 Geometric Sequence

In a geometric sequence each next element is computed by multiplying the previous element by a constant:

$$R_{n+1} = qR_n. \tag{9.8}$$

Consider a sequence with the first element R_1. Its n-th element is given by

$$R_n = q^{n-1}R_1. \tag{9.9}$$

Theorem 9.2 *The sum S of all elements of a geometric sequence is given by*

$$S = \sum_{n=1}^{N} R_n = R_1 \frac{1-q^N}{1-q}. \tag{9.10}$$

Below are two proofs of this theorem. A third proof is the subject of exercise 2 for this chapter.

Proof

The classic and most elegant proof is as follows. Let's multiply the sum S of the elements by q and use the explicit values of elements $R_n = q^{n-1}R_1$ in the resulting expression:

$$\begin{aligned} qS &= qR_1 + qR_2 + \ldots + qR_{N-1} + qR_N \\ &= qR_1 + q^2R_1 + \ldots + q^{N-1}R_1 + q^N R_1. \end{aligned} \tag{9.11}$$

In the right-hand side, we add and subtract R_1:

$$qS = R_1 + qR_1 + q^2R_1 + \ldots + q^{N-1}R_1 + q^N R_1 - R_1. \tag{9.12}$$

Now in the right-hand side we recognize the original sum S of all elements, complemented with two additional terms:

$$qS = S + q^N R_1 - R_1. \tag{9.13}$$

This is an equation for the desired sum of elements of a geometric sequence. We solve it for S to get formula (9.10).

The second proof shows that we can use induction here.

Proof

As usual, we consider the case $N = 1$ and then prove that if the theorem is true for some value of N, then it is true for $N + 1$.

1. *For the series with just one element, we have $N = 1$, and the sum is equal to the first and only element, R_1. Equation (9.10) works:*

$$\begin{aligned}\sum_{n=1}^{N} R_n &= R_1 \frac{1-q^N}{1-q} \\ &= R_1 \frac{1-q}{1-q} \\ &= R_1.\end{aligned} \tag{9.14}$$

2. *For the series that contains $N+1$ elements, we get:*

$$\begin{aligned}\sum_{n=1}^{N+1} R_n &= \sum_{n=1}^{N} R_n + R_1 q^N \\ &= R_1 \frac{1-q^N}{1-q} + R_1 q^N \\ &= R_1 \frac{1-q^N + q^N(1-q)}{1-q} \\ &= R_1 \frac{1-q^{N+1}}{1-q},\end{aligned} \tag{9.15}$$

which completes the proof.

For example,

$$\frac{1}{2} + \frac{1}{2^2} + \frac{1}{2^3} + \ldots + \frac{1}{2^N} = \frac{1}{2} \cdot \frac{1 - \frac{1}{2^N}}{1 - \frac{1}{2}} = 1 - \frac{1}{2^N}. \tag{9.16}$$

Note that this sum is less than 1 for any N. We will use this fact in section 9.12.

9.3 Infinite Sequences

Consider a sequence of numbers that extends indefinitely. Some examples include (here we assume that the obvious pattern in each sequence continues):

A. $1, \frac{1}{2}, \frac{1}{3}, \frac{1}{4}, \ldots$

B. $\pi, -\pi, \pi, -\pi, \pi, \ldots$

C. $\sin(1), \sin(2), \sin(3), \sin(4), \ldots$

Sometimes, a sequence is defined using sums or other procedures. For example, we can define the following sequence:

$$1; 1+\frac{1}{2}; 1+\frac{1}{2}+\frac{1}{3}; 1+\frac{1}{2}+\frac{1}{3}+\frac{1}{4}; ... \quad (9.17)$$

Another way to define a sequence is to do it recursively. For example, the Fibonacci sequence starts from 1, 1; then each next term is the sum of the two previous terms:

$$1, 1, 2, 3, 5, 8, 13, 21, ... \quad (9.18)$$

This sequence has many interesting properties that we will investigate in chapter 10.

9.4 Limits: Definition

Sometimes we can detect that values in a sequence approach a particular value and "stay there." For example, consider the sequence $1, \frac{1}{2}, \frac{1}{3}, \frac{1}{4}, ...$. The values get closer and closer to zero. This situation is formalized by the notion of a *limit* of a sequence. The definition of a limit is as follows:

Consider an infinite sequence $a_1, a_2, a_3, a_4, ...$ We say that this sequence has limit A (or *converges* to A) if for any positive value d there exists a number N such that for any $n > N$ it is the case that $|a_n - A| < d$.

Let's dissect this definition and see how it works. First, condition $|a_n - A| < d$ means that element a_n of the sequence differs from A by less than d. Note that a_n can be greater than A, less than A, or equal to A, but the absolute value of the difference between a_n and A is less than some value d (see figure 9.1).

The definition of a converging sequence means that for any positive value of d, no matter how large or small, all elements of the sequence starting from that element number N will differ from A by less than d. Take, for example, the sequence $a_1 = 1 + \frac{1}{1}; a_2 = 1 + \frac{1}{2}; a_3 = 1 + \frac{1}{3}; a_4 = 1 + \frac{1}{4}....$ Intuitively, we know that it gets ever closer to 1. We assume that it converges to $A = 1$. Let's consider examples of how for any positive value d all elements of the sequence, starting from some specific element, are within distance d from 1:

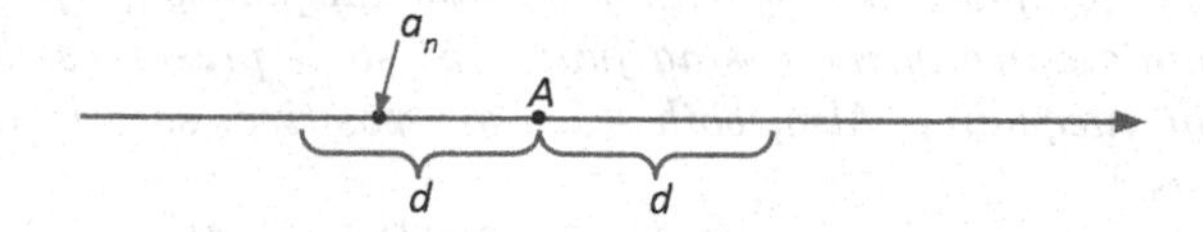

FIGURE 9.1
A vicinity of the convergence value on a number line

1. Take $d = 0.1$. There must be such value of N, that all elements of the sequence after number N differ from limit A by less than d. Indeed, look at element $a_{11} = 1+\frac{1}{11}$. It differs from the convergence limit $A = 1$ by $\frac{1}{11}$, which is less than d. Moreover, all subsequent elements are even closer to A.
2. Take $d = 0.01$. Similarly, element $a_{101} = 1 + \frac{1}{101}$ differs from A by less than d. All subsequent elements are even closer to A.
3. Take $d = \pi$. Then $N = 1$. All elements starting from the first one differ from A by less than d.

We see that in all three examples there exist such a number N that all elements of the sequence starting from N differ from A by less than d. These observations can be formalized and extended to all possible values of $d > 0$.

If a sequence has a limit, we say that it is *convergent*. If there is no such number that satisfies the above definition of a limit, we say that the sequence is *divergent*.

9.5 Does the Geometric Sequence Converge?

The convergence of the geometric sequence given by $R_{n+1} = qR_n$ depends on the value of its multiplier q:

Theorem 9.3 *The geometric sequence given by $R_{n+1} = qR_n$ converges to zero if $|q| < 1$; to R_1 if $q = 1$; and it does not converge for other values of q.*

Proof

1. *Consider $|q| < 1$. We look at two sub-cases.*

   (a) *The case of $|q| < 1$ and $q \neq 0$. (A similar proof is the subject of exercise 1 for chapter 6.) The n-th element of the sequence differs from zero by the following amount: $|R_1 q^{n-1} - 0| = |R_1 q^{n-1}| = |R_1||q|^{n-1}$. Let's find an N such that for all $n > N$ we have*

$$|R_1||q|^{n-1} < d. \tag{9.19}$$

We compute the logarithm of this inequality (logarithm is a monotonically increasing function, so it preserves the sign of the inequality. Also, both sides are positive, so a logarithm exists.)

$$(n-1)\ln|q| + \ln|R_1| < \ln d. \tag{9.20}$$

Next, we subtract $\ln|R_1|$ and divide the resulting inequality by $\ln|q|$. Since $|q| < 1$, the logarithm is negative, and we need to *flip the sign of the inequality. We get*

$$n > \frac{\ln d - \ln |R_1|}{\ln |q|} + 1. \tag{9.21}$$

Note that we did not place any assumptions on the value of d, except assuming $d > 0$. It can be small (then $\ln d < 0$) or large (then $\ln d > 0$). We see that for any positive value of d, there is such a threshold, that for all values of n above that threshold the corresponding elements of the geometric sequence differ from zero by less than d. Therefore, the sequence converges to zero.

(b) *The case of $q = 0$. All elements, except, possibly, R_1 are zeroes. The sequence therefore converges to zero.*

2. *Consider $q = 1$. All elements are equal to R_1. The sequence therefore converges to that value.*
3. *Consider all other values of q. We look at two subcases:*

   (a) *Case $q = -1$. The sequence oscillates between R_1 and $-R_1$. We see that it does not converge.*

   (b) *Case $|q| > 1$. The geometric sequence diverges in this case. A proof of this fact is the subject of exercise 21 for this chapter.*

9.6 Arithmetic Operations for Sequences

Suppose that we have two sequences: $a_1, a_2, a_3, a_4, \ldots$ and $b_1, b_2, b_3, b_4, \ldots$. Let's form a third sequence: $a_1 + b_1, a_2 + b_2, a_3 + b_3, a_4 + b_4, \ldots$

Theorem 9.4 *If sequence a_n converges to A and sequence b_n converges to B, then the element-wise sum of these sequences converges to $A + B$.*

Proof

For any positive value d we must prove that there exists some value of N such that $|a_n + b_n - (A + B)| < d$ for all $n > N$. We regroup the terms:

$$|(a_n - A) + (b_n - B)| < d. \tag{9.22}$$

Next, we apply theorem 3.11 to get:

$$|(a_n - A) + (b_n - B)| \le |(a_n - A)| + |(b_n - B)|. \tag{9.23}$$

It will be sufficient to prove that

$$|(a_n - A)| + |(b_n - B)| < d. \tag{9.24}$$

This condition will be satisfied if we require that

$$\begin{aligned} |a_n - A| &< \frac{d}{2}, \\ |b_n - B| &< \frac{d}{2}. \end{aligned} \tag{9.25}$$

Note that there are other ways to satisfy inequality (9.24), but splitting it in two halves works. Since both sequences converge, there exist numbers K and M such that

$$\begin{aligned} |a_n - A| &< \frac{d}{2} \textit{ for } n > K, \\ |b_n - B| &< \frac{d}{2} \textit{ for } n > M. \end{aligned} \tag{9.26}$$

These two numbers K and M may be different. We select the larger of the two: $N = \max(M, K)$. Then for all $n > N$ we guarantee both conditions (9.26). In turn, they guarantee condition (9.24), which implies (9.22). This proves the convergence.

For example, consider two sequences: $1, \frac{1}{2}, \frac{1}{3}, \frac{1}{4}, \ldots$ and $\pi, \frac{\pi}{2}, \frac{\pi}{3}, \frac{\pi}{4}, \ldots$. Both converge to zero. Their sum is converging to $1 + \pi, \frac{1+\pi}{2}, \frac{1+\pi}{3}, \frac{1+\pi}{4}, \ldots$. We see that it also converges to zero.

The next two theorems deal with a sum or a product of a sequence and a constant:

Theorem 9.5 *If sequence a_n converges to A, and C is a constant, then sequence $C + a_n$ converges to $C + A$.*

Theorem 9.6 *If sequence a_n converges to A, and C is a constant, then sequence Ca_n converges to CA.*

A proof of theorems 9.5 and 9.6 is the subject of exercise 19 for this chapter. For the product of two sequences there is a theorem that is similar to theorem 9.4:

Theorem 9.7 *The element-wise product of two convergent sequences converges to the product of their limits. In other words, if $a_n \to A$ and $b_n \to B$, then $a_n b_n \to AB$.*

Proof

We define numbers ε_n and δ_n as follows:

$$\begin{aligned} \varepsilon_n &= a_n - A, \\ \delta_n &= b_n - B. \end{aligned} \tag{9.27}$$

We see that sequences ε_n and δ_n converge to 0. The n-th element of sequence $a_n b_n$ can be expressed as

$$a_n b_n = (A + \varepsilon_n)(B + \delta_n). \tag{9.28}$$

Let's compute the difference between this element and AB:

$$\begin{aligned} a_n b_n - AB &= (A + \varepsilon_n)(B + \delta_n) \\ &= A\delta_n + B\varepsilon_n + \varepsilon_n \delta_n. \end{aligned} \tag{9.29}$$

We need to prove that for any positive d there is such a value of N that for all $n > N$ it is the case that

$$|A\delta_n + B\varepsilon_n + \varepsilon_n \delta_n| < d. \tag{9.30}$$

From theorem 3.12 we know that

$$|A\delta_n + B\varepsilon_n + \varepsilon_n| \leq |A||\delta_n| + |B||\varepsilon_n| + |\varepsilon_n||\delta_n|. \tag{9.31}$$

So far, we obtained an expression in the right-hand side that also contains an element-wise product $|\varepsilon_n||\delta_n|$ of two sequences. Our plan is to use theorem 9.6 to deal with the right-hand side of inequality (9.31). To do that, we need to have a sequence that is multiplied by a constant, and not by another sequence. Let's replace $|\varepsilon_n|$ in that product with the maximum value $\max_n(|\varepsilon_n|)$ of all elements. Since sequence $|\varepsilon_n|$ converges, such maximum value exists.

$$|\varepsilon_n||\delta_n| \leq |\delta_n| \max_n(|\varepsilon_n|). \tag{9.32}$$

The value of $\max_n(|\varepsilon_n|)$ is no longer a sequence, it is just a constant. After replacing $|\varepsilon_n|$ with the maximum value, inequality (9.31) still holds:

$$|A\delta_n + B\varepsilon_n + \varepsilon_n| \leq |A||\delta_n| + |B||\varepsilon_n| + |\delta_n| \max_n(|\varepsilon_n|). \tag{9.33}$$

From equations (9.27) we know that sequences ε_n and δ_n converge to zero. We use theorem 9.6 to conclude that there exist such element numbers N_1, N_2, and N_3 that

$$\begin{aligned} |A||\delta_n| &< \frac{d}{3} \text{ for } n > N_1, \\ |B||\varepsilon_n| &< \frac{d}{3} \text{ for } n > N_2, \\ \max(|\varepsilon_n|)|\delta_n| &< \frac{d}{3} \text{ for } n > N_3. \end{aligned} \tag{9.34}$$

Then for all $n > \max(N_1, N_2, N_3)$ all three inequalities in (9.34) will be satisfied, and we will have

$$|A\delta_n + B\varepsilon_n + \delta_n \varepsilon_n| < d, \tag{9.35}$$

which proves the theorem.

The opposite is not true: if $a_n b_n \to AB$, there is no guarantee that a_n and b_n converge at all. Similarly, if $a_n + b_n \to A + B$, there is no guarantee that a_n and b_n converge. For example consider sequences $a_n = 1, -1, 1, -1, \ldots$ and $b_n = -1, 1, -1, 1, \ldots$. Neither of these sequences converge. However, both their pairwise product $a_n b_n = -1, -1, -1, \ldots$ and their sum $a_n + b_n = 0, 0, 0, \ldots$ converge!

Theorem 9.8 *If sequence a_n converges to A and sequence b_n, where $b_n \neq 0$ for all n, converges to B such that $B \neq 0$, then a_n/b_n converges to A/B.*

Proof

From the previous theorems we conclude that if sequences a_n and b_n converge, then sequence $a_nB - b_nA$ also converges (see theorems 9.6 and 9.4). This means that for any positive d we have[1]

$$|a_nB - b_nA| < \frac{B^2}{2}d \tag{9.36}$$

for all n that are greater than some value N. Since $b_n \to B$, we have $|b_n - B| < |B/2|$ starting from some number $n = M$. This means that $|b_n| > |B/2|$ for all n that are greater than some value M. Then

$$|a_nB - b_nA| < \frac{B^2}{2}d < |b_nB|d. \tag{9.37}$$

We divide this inequality by $|b_nB|$ to get:

$$\left|\frac{a_nB - b_nA}{b_nB}\right| < d. \tag{9.38}$$

This is equivalent to

$$\left|\frac{a_n}{b_n} - \frac{A}{B}\right| < d, \tag{9.39}$$

which is the condition for convergence of sequence a_n/b_n to A/B.

9.7 Monotone and Bounded Sequences

Here, we define three important terms for sequences.

1. One concept that we have defined for functions becomes particularly fruitful for sequences. We say that a sequence is strictly monotonically increasing if $S_{n+1} > S_n$, and strictly monotonically decreasing if $S_{n+1} < S_n$ for all n. For example:
   (a) Sequence $1, \frac{1}{2}, \frac{1}{4}, \frac{1}{8}, \frac{1}{16}, ...$ is strictly monotonically decreasing.
   (b) Sequence $-1, -\frac{1}{2}, -\frac{1}{4}, -\frac{1}{8}, -\frac{1}{16}, ...$ is strictly monotonically increasing.
   (c) Sequence $1, -\frac{1}{2}, \frac{1}{4}, -\frac{1}{8}, \frac{1}{16}, ...$ is not monotonic.

[1] In the right-hand side, the familiar positive d is replaced by $B^2d/2$, but that latter value still conforms to the definition of a limit, which requires the right hand side to be an arbitrary positive number.

2. A bounded sequence is such that all its terms lie within a finite interval. Mathematically, a sequence $a_1, a_2, a_3, \ldots$ is bounded if there exist numbers b and c, such that $b < a_n < c$ for all n. For example:
   (a) Sequence $1, \frac{1}{2}, \frac{1}{4}, \frac{1}{8}, \frac{1}{16}, \ldots$ is bounded.
   (b) Sequence $1, 2, 3, 4, 5, \ldots$ is not bounded.
3. For some sequence $a_1, a_2, a_3, \ldots$, a subsequence is another infinite sequence that is obtained from the original sequence by picking some of its elements, while keeping their order. For example, let's take sequence $1, 2, 3, 4, 5, \ldots$. We can form many subsequences from it, such as:
   (a) $1, 3, 5, 7, \ldots$ (odd numbers selected),
   (b) $2, 3, 5, 297, 1001, 1002, 1003, \ldots$ (some elements from the first thousand are missing),
   (c) $1, 2, 3, 4, \ldots$ (all elements are present).

   If we have the original sequence $a_1, a_2, a_3, \ldots$, a subsequence is denoted as $a_{n_1}, a_{n_2}, a_{n_3}, a_{n_4}, \ldots$, where $n_1, n_2, n_3, \ldots$ are positive integers such that $n_1 < n_2 < n_3 < n_4 < \ldots$.

There are several important theorems that apply to monotonic and bounded sequences. The one below is quite intuitive: if a sequence grows but is bounded from above, it must converge to some limit as it "hits the ceiling." Same is true for a decreasing sequence that is bounded from below – it "hits the floor":

Theorem 9.9 *A bounded monotonic sequence is convergent.*

Proof

We prove this theorem for monotonically increasing sequences. Consider the set of numbers $\{a_n\}$ that contains all elements of the sequence. In section 2.8.7 we postulated that this set must have the least upper bound c. Due to the definition of the least upper bound, for any $d > 0$ there is some number K, such that $c - a_K < d$. Since the sequence is increasing, it is the case that $a_k > a_K$ for any $k > K$. Then $c - a_k < d$ for any $k \geq K$, and we also remember that a_k does not exceed c. This fits the definition of the sequence converging to c.

The proof for monotonically decreasing sequences is analogous.

Section 6.4 heuristically introduced the notion of real-valued exponents. Now, we can define real exponents in a more meaningful way using theorem 9.9. Let's consider a sequence of numbers 2^3, $2^{3.1}$, $2^{3.14}$, $2^{3.141}$, $2^{3.1415}$, $2^{3.14159}, \ldots$, which is formed by raising 2 to rational powers that are successive truncated decimal representations of π. We know that the exponent function for rational powers is monotonically increasing (see theorem 6.12). We also understand that this sequence is bounded (indeed, by the same monotonic property all elements are smaller than 2^4). Therefore, this sequence has a limit. We define 2^π as that limit.

Theorem 9.10 *If a sequence is convergent (has a limit), than any its subsequence is convergent and has the same limit.*

Proof

By definition, if a sequence $a_1, a_2, a_3, ...$ converges to A, then for every positive d there exist such number N that $|a_n - A| < d$ for all $n > N$. Suppose we have a subsequence $a_{n_1}, a_{n_2}, a_{n_3}, a_{n_4}, ...$ We just need to find the first element a_{n_k} such that $n_k > N$. Since all elements of the subsequence are in order, all following elements will correspond to elements of the original sequence with larger indices, and therefore will satisfy the convergence criterion: $|a_{n_m} - A| < d$ for all $m > k$.

Theorem 9.11 *If a sequence is bounded, than any of its subsequences is bounded (but not necessarily by the same interval).*

Proof

If any element of a sequence lies between two numbers b and c, and every element of the subsequence is an element of the original sequence, then every element of the subsequence also lies between the same two numbers.

However, keep in mind that the bound for the original sequence may not be as tight for a subsequence. For example, take sequence $-1, 1, -1, 1, -1, 1, \ldots$ It is bounded by -1 and 1. However this bound is not tight for a subsequence $-1, -1, -1, \ldots$ that contains only odd elements of the original sequence.

Theorem 9.12 *If two sequences a_n and c_n converge to the same limit and if there is another sequence b_n such that $a_n \leq b_n \leq c_n$, then this sequence b_n converges to the same limit. (This is known as the squeeze theorem.)*

Proof

Suppose sequences a_n and c_n converge to the same limit A. We consider some arbitrary positive value d. Due to the definition of convergence, this means that there exist numbers K and M such that $|A - a_k| < d$ for all $k \geq K$ and $|A - c_m| < d$ for all $m \geq M$. Consider the greater of the values of K and M:

$$J = \max(K, M). \tag{9.40}$$

Then for all $j \geq J$ it is the case that

$$\begin{aligned} |A - a_j| &< d, \\ |A - c_j| &< d. \end{aligned} \tag{9.41}$$

From this we obtain

$$\begin{aligned} a_j &> A - d, \\ c_j &< A + d \end{aligned} \tag{9.42}$$

for all $j > J$. *Since* $a_n \leq b_n \leq c_n$, *the last two inequalities also mean that*

$$\begin{aligned} b_j &> A - d, \\ b_j &< A + d, \end{aligned} \tag{9.43}$$

which is equivalent to

$$|A - b_j| < d \tag{9.44}$$

for all $j > J$. *This is the definition of sequence* b_j *converging to* A.

9.8 The Bolzano-Weierstrass Theorem

Here is a statement of this weird but powerful theorem:

Theorem 9.13 *Every bounded sequence has a convergent subsequence.*

Let's think about this a bit. We already know that a bounded monotonic sequence is always convergent. We also know that in a convergent sequence, any subsequence converges to the same limit. So, if the bounded sequence in question is monotonic, then the theorem works.

However, the theorem is formulated for any bounded sequence, not necessarily monotonic. Take a sequence like $-1, 1, -1, 1, -1, ...$ Yes, it does have convergent subsequences, if the elements of these subsequences are selected exclusively from the odd elements or from the even elements of the original sequence, but not from both these classes. These subsequences converge correspondingly to -1 and 1. However, there is a gazillion other bounded sequences. Take, for example this one:

1. We use the decimal representation of $\pi = 3.14159265358979323846264...$
2. For the first element of this sequence, we take the first ten digits after the decimal point and square them: $a_1 = 1415926535^2 = 2004847952517106225$.
3. For the second element of this sequence, we take the next ten digits after the decimal point and square them: $a_2 = 8979323846^2 = 80628256731344231716$, and so on...

This sequence is bounded. Its smallest possible value is equal to 0 (if all ten digits happen to be zeroes). Its largest possible value is $9999999999^2 = 99999999980000000001$. The Bolzano-Weierstrass theorem states that there is a subsequence within this crazy sequence that converges to a finite value. But it does not tell us what that subsequence is!

Here is another example: take sequence $\sin(1), \sin(2), \sin(3), \sin(4), ...$ This sequence jumps all over between values of -1 and 1. Therefore, it is bounded. Then, it must contain a convergent subsequence!

This theorem has more than one proof. Here is one of them.

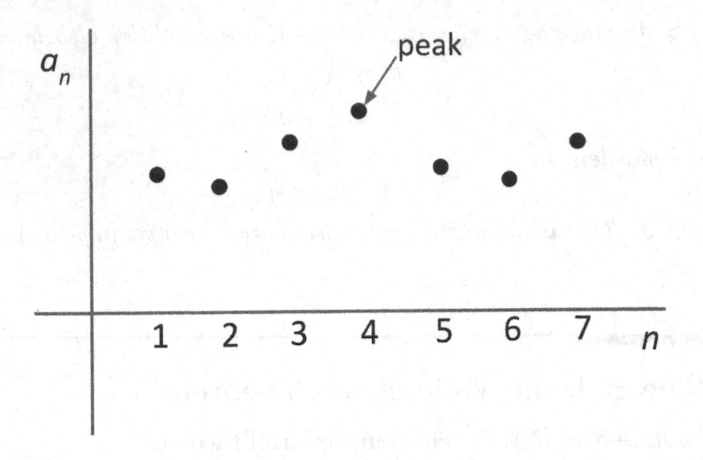

FIGURE 9.2
A peak of a sequence

Proof
First, we define a peak in the sequence. Consider a sequence $a_1, a_2, a_3, a_4, a_5, \ldots$ *Then suppose that element* K *has the property that all subsequent elements of this sequence are less than* a_K. *Mathematically, this is written as follows: for all* $n > K$ *it is the case that* $a_n < a_K$. *We call this element a peak. This is illustrated in figure 9.2, where element 4 is a peak of a sequence: all elements after element 4 are smaller (obviously, we can show only a few of them on the plot).*

Next, we consider two possibilities:

1. *The sequence has an infinite number of peaks. Here are two examples when a sequence has an infinite number of peaks:*
   (a) *Sequence* $1, \frac{1}{2}, \frac{1}{4}, \frac{1}{8}, \frac{1}{16}, \ldots$ *For each element, all subsequent elements are smaller. Therefore, each element is a peak.*
   (b) *Sequence* $1, \frac{3}{2}, \frac{1}{4}, \frac{3}{8}, \frac{1}{16}, \frac{3}{32}, \ldots$ *Here all even elements are peaks, but odd elements are not peaks.*

   If a sequence has an infinite number of peaks, let us select all these peaks as a subsequence. For this subsequence, each next peak must be lower than the previous one, or that previous peak would not be a peak. Therefore, all these peaks form a monotonically decreasing subsequence. Since the original sequence is bounded, this subsequence is bounded too. We get a bounded monotonic (sub)sequence, which means that it is convergent.

2. *The sequence has a finite number of peaks (or no peaks at all). Take the element right after the last peak (or the first element if there are no peaks). We denote this element as* a_L. *This element*

is not a peak (remember, it comes after the last peak!). Therefore, there must be another element somewhere further in the sequence, which is greater than or equal to a_L (otherwise, a_L would be a peak). We denote that new element a_M (note that $M > L$). Since a_M too comes after the last peak, this new element is not a peak either. Then there must be yet another element even further in the sequence that is greater than or equal to a_M; we denote it a_N. We can continue this procedure indefinitely. It produces an infinite subsequence of elements $a_L, a_M, a_N, ...$, each greater than or equal to the previous one. This is a monotonically increasing bounded sequence, which means that it has a limit.

This is a truly beautiful proof of existence, in part because it is not a constructive proof. We proved that a convergent subsequence exists without pointing to it.

Let's go back to the example of a sequence $\{\sin(n)\}$, where $n = 1, 2, 3, \ldots$. According to the Bolzano-Weierstrass theorem, it has a convergent subsequence. Turns out, there is more than one convergent subsequence. Here is a proof of that. Let us denote a convergent subsequence $\sin(n_1), \sin(n_2), \sin(n_3), \sin(n_4), ...$, where $n_1, n_2, n_3, n_4, ...$ are positive integers (they are simply element numbers of the original sequence) such that $n_1 < n_2 < n_3 < n_4 < ....$ Let us assume that this subsequence converges to some value A.

Now let's form a new subsequence with element numbers $n_1 + M, n_2 + M, n_3 + M, n_4 + M, ...$, where M is some positive integer value. We consider the value of a k-th element in this new subsequence:[2]

$$\begin{aligned} \sin(n_k + M) &= \sin n_k \cos M + \cos n_k \sin M \\ &= \sin n_k \cos M + \sqrt{1 - \sin^2 n_k} \sin M. \end{aligned} \tag{9.45}$$

In the right-hand side, we get $\sin n_k \to A$. Then

$$\sin(n_k + M) \to A \cos M + \sqrt{1 - A^2} \sin M. \tag{9.46}$$

We see that if we shift the original sequence by some arbitrary number of elements M, we get a new convergent subsequence. Since M is arbitrary, we can get more than one new convergent subsequences this way.

9.9 More on the Ratio of Two Sequences

We already investigated the element-wise ratio $\frac{a_1}{b_1}, \frac{a_2}{b_2}, \frac{a_3}{b_3}, \frac{a_4}{b_4}, ...$ of two convergent sequences a_n and b_n, where $b_n \neq 0$ for all n. If b_n converges to a nonzero limit, then the ratio of these two sequences converges to the ratio of limits (theorem 9.8). As we might assume, the situation changes if $b_n \to 0$.

[2]The following equation has a small flaw, fixing which is a subject of exercise 22 for this chapter.

If a_n in the numerator converges to a finite (nonzero) limit, the resulting sequence becomes unbounded and does not converge. For example, consider sequences $1, 1, 1, 1, ...$ and $\frac{1}{2}, \frac{1}{3}, \frac{1}{4}, ...$ For their ratio, we get:

$$\frac{1}{\frac{1}{2}}, \frac{1}{\frac{1}{3}}, \frac{1}{\frac{1}{4}}, \frac{1}{\frac{1}{5}}, ... = 2, 3, 4, 5, ... \tag{9.47}$$

which obviously does not converge.

However if *both* the numerator and the denominator sequences converge to zero, all bets are off: the ratio may converge to zero, to a finite value, or diverge. In addition, if both sequences diverge, their ratio may still converge (to zero or a finite value).

Let's consider an example of a ratio of two diverging sequences: $c_n = q^n/n!$, where q is some positive number, such that $q > 1$. This sequence can be viewed as a ratio of two sequences: $a_n = q^n$ and $b_n = n!$. Both these sequences diverge (grow indefinitely). What can we say about the ratio? First, we need to determine if it converges at all.

All elements of this sequence are positive, so it is bounded from below (by zero). Next, we will write this sequence in the recursive form, which means that we express the $(n+1)$-th element through the n-th one. Since $(n+1)! = n! \cdot (n+1)$, and $q^{n+1} = q^n \cdot q$, we get:

$$c_{n+1} = \frac{q}{n+1} c_n. \tag{9.48}$$

Let's look closely at this equation. For values of $n+1$ that are larger than q, each next element is produced by multiplying the previous element by a factor that is smaller than 1. Therefore, starting from some $N+1 > q$, each next element is smaller than the previous, and the sequence is monotonically decreasing. Therefore, it is monotonically decreasing (at least starting from a certain element) *and* bounded, which means that it converges.

This conclusion opens a path to compute the limit. Let's look again at equation (9.48). We know that c_n converges; we denote its limit as C. Obviously, sequence c_{n+1} in the left-hand side converges to the same limit: $c_{n+1} \to C$. The quotient $q/(n+1)$ itself is a sequence, which converges to zero. Since all sequences in that equation converge, we can use the theorem for the limit of a product:

$$C = 0 \cdot C. \tag{9.49}$$

The only solution for this equation is $C = 0$, which means that the original sequence converges to zero.

This is an example of an *indeterminate form*, when the two sequences in an expression do not easily yield a limit for this expression. Their are different types of indeterminate forms. The sequence $c_n = q^n/n!$ is of the $\frac{\infty}{\infty}$ type. In addition to $\frac{\infty}{\infty}$, other types of indeterminate forms are:

$0 \cdot \infty; 1^\infty; 0^0; \infty^0; \infty - \infty$

None of these notations means what it says: we cannot raise a zero to zero power, nor we can subtract one infinity from another. These notations just show the limits of the two sequences that are combined in a particular mathematical expression. For all these indeterminate forms, the limit may or may not exist; if it does exist, it may be zero or a finite value. Each case must be considered separately.

9.10 A Sequence with Nested Radicals

Let's consider the following sequence:

$$\begin{aligned} a_1 &= \sqrt{c}, \\ a_2 &= \sqrt{c+\sqrt{c}}, \\ a_3 &= \sqrt{c+\sqrt{c+\sqrt{c}}}, \\ a_4 &= \sqrt{c+\sqrt{c+\sqrt{c+\sqrt{c}}}}, \\ a_5 &= ..., \end{aligned} \tag{9.50}$$

where $c > 0$. This sequence is produced by a recursive scheme

$$a_{n+1} = \sqrt{c + a_n}. \tag{9.51}$$

Does it have a limit? Let's see if this sequence is monotonic and bounded.

Theorem 9.14 *Sequence (9.50) is monotonically increasing.*

Proof

We prove this theorem by induction.

1. *We square inequality*

$$a_2 > a_1 \tag{9.52}$$

to get

$$c + \sqrt{c} > c, \tag{9.53}$$

which is true.

2. *Consider now*

$$a_n > a_{n-1}. \tag{9.54}$$

We add c to both sides of this inequality and extract the square root. According to equation (9.51) we obtain

$$a_{n+1} > a_n, \tag{9.55}$$

which completes the proof.

Now let's prove that it is also bounded.

Theorem 9.15 *Sequence (9.50) is bounded.*

Proof
We see that zero can serve as a lower bound, but we need to prove that the upper bound exists. Let's start from the already established fact that this sequence is monotonic:

$$a_{n+1} > a_n. \tag{9.56}$$

We use equation (9.51) to get:

$$\sqrt{c + a_n} > a_n. \tag{9.57}$$

Both sides are positive; therefore, we can square this inequality and keep the sign:

$$c + a_n > a_n^2. \tag{9.58}$$

This is equivalent to

$$a_n^2 - a_n < c. \tag{9.59}$$

The fact that a_n *is bounded is proved by contradiction. Let's assume that* $a_n > 1 + c$ *for some value of* n*. Then, we have* $a_n - 1 > 0$ *and*

$$\begin{aligned} a_n^2 - a_n &= a_n(a_n - 1) \\ &> (1 + c)c \\ &= c + c^2. \end{aligned} \tag{9.60}$$

Then from inequalities (9.59) and (9.60) we obtain $c + c^2 < c$*, which is false.*

Now, we know that this sequence is monotonic and bounded, which means that it is convergent. We square equation (9.51) to get $c + a_n = a_{n+1}^2$ and set values of a_{n+1} and a_n to their limit, which we denote A. If $a_{n+1} \to A$ and $a_n \to A$, then

$$c + A = A^2. \tag{9.61}$$

This is a quadratic equation for A. Its solutions are:

$$A_{1,2} = \frac{1 \pm \sqrt{1 + 4c}}{2}. \tag{9.62}$$

Since all elements of the sequence are positive, the limit must be nonnegative, which means that we have to select the plus sign in the above solution:

$$A = \frac{1 + \sqrt{1 + 4c}}{2}. \tag{9.63}$$

Let's consider two special cases that we will use later:

1. For $c = 1$ we get

$$A_1 = \frac{1 + \sqrt{5}}{2} \approx 1.61803398874989484 82. \tag{9.64}$$

   We will encounter this number in chapter 10.

2. For $c = 2$ we get

$$A_2 = \frac{1+\sqrt{9}}{2} = 2. \tag{9.65}$$

We will use this result in section 9.11.

9.11 More Sequences with Nested Radicals

In this section we derive two approximations for π. We will do this from the definition of π—that is, from the ratio of the circumference of a circle to its diameter. For the circumference, we consider a sequence of inscribed polygons with ever increasing number of sides. We will determine the perimeter of each polygon, and the circumference of the circle will be then computed from a limit of these perimeters as the number of sides goes to infinity.

Figure 9.3 shows a circle and an inscribed hexagon. Consider side DB. We draw radii CD and CB to the ends of this side. For simplicity, we consider a unit circle, so the lengths of the radii are equal to 1. The circumference of such a circle is equal to 2π. Since we are interested in the value of π, we will seek the length of a half of a circle. That means that we will consider a half of the perimeter of the polygons that are inscribed in the circle.

The polygon has n sides. The measure of the half-angle GCB is denoted as α_n. Then the length of a half-side AB of the polygon is equal to $L_n = \sin \alpha_n$, and a half-perimeter of the polygon is given by $n \sin \alpha_n$.

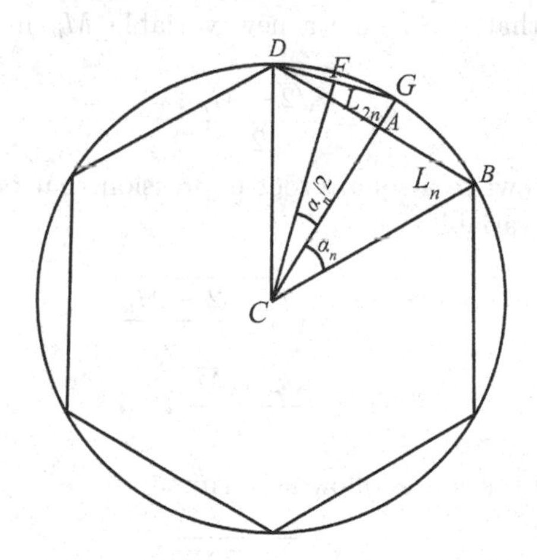

FIGURE 9.3
A circle with an inscribed hexagon

Our strategy is to double the number of sides of this polygon again and again. A side of the polygon with $2n$ sides is shown as DG. The length of a half-side of this new polygon is equal to $L_{2n} = \sin \alpha_n/2$, and a half-perimeter of that polygon is given by $2n \sin \alpha_n/2$. Our goal is to express the half-perimeter of the polygon with $2n$ sides through the half-perimeter of the polygon with n sides and then continue doubling the number of sides to get successive approximations for π. To do that, we need a recursive formula that expresses L_{2n} through L_n. Since we know that angle α_n is halved, we start from the trigonometric identity for a half-angle:

$$\sin \frac{\alpha}{2} = \sqrt{\frac{1 - \cos \alpha}{2}}. \tag{9.66}$$

Since the half-perimeter of the n-polygon uses sine, and not cosine, of α_n, we want to use a sine in the right-hand side. That way we will connect the sines of the angles for polygons with n and $2n$ sides. We recall the Pythagoras identity to get

$$\sin \frac{\alpha}{2} = \sqrt{\frac{1 - \sqrt{1 - \sin^2 \alpha}}{2}}. \tag{9.67}$$

This formula can already be used to connect L_{2n} and L_n. The recursion is now given by

$$L_{2n} = \sqrt{\frac{1 - \sqrt{1 - L_n^2}}{2}}. \tag{9.68}$$

However, we can make this recursive formula more apparent. We will also connect it to the type of sequences with nested radicals that we explored in section 9.10. To do that, we define a new variable M_n using the following equation:

$$L_n = \frac{\sqrt{2 - M_n}}{2} \tag{9.69}$$

for any n. Next, we rewrite a square root expression that occurs in equation (9.68) using the new variable:

$$\begin{aligned} \sqrt{1 - L_n^2} &= \sqrt{1 - \frac{2 - M_n}{4}} \\ &= \frac{\sqrt{2 + M_n}}{2}. \end{aligned} \tag{9.70}$$

Then equation (9.68) takes the following form:

$$\begin{aligned} L_{2n} &= \sqrt{\frac{1 - \frac{\sqrt{2+M_n}}{2}}{2}} \\ &= \frac{\sqrt{2 - \sqrt{2 + M_n}}}{2}. \end{aligned} \tag{9.71}$$

On the other hand, we can apply the definition of M_n to produce

$$L_{2n} = \frac{\sqrt{2 - M_{2n}}}{2}. \tag{9.72}$$

Let's compare equations (9.71) and (9.72). Equation (9.72) contains M_{2n} in the place where equation (9.71) contains $\sqrt{2 + M_n}$. In order for these two equations to produce the same value L_{2n} we must have

$$M_{2n} = \sqrt{2 + M_n}, \tag{9.73}$$

which means that sequence M_n can be computed recursively in its own way. That is, as we double the number of sides, we replace the current value of M_n with $M_{2n} = \sqrt{2 + M_n}$. The half P_n of the perimeter of a polygon with n sides would then be given by

$$\begin{aligned} P_n &= nL_n \\ &= \frac{1}{2}n\sqrt{2 - M_n}. \end{aligned} \tag{9.74}$$

Since $M_{2n} = \sqrt{2 + M_n}$, the half-perimeter of the $2n$-polygon is then given by

$$P_{2n} = \frac{1}{2}2n\sqrt{2 - \sqrt{2 + M_n}}. \tag{9.75}$$

Continuing this process of doubling the number of sides, we will get

$$\begin{aligned} P_{4n} &= \frac{1}{2}4n\sqrt{2 - \sqrt{2 + \sqrt{2 + M_n}}}, \\ P_{8n} &= \frac{1}{2}8n\sqrt{2 - \sqrt{2 + \sqrt{2 + \sqrt{2 + M_n}}}}, \\ &\ldots \end{aligned} \tag{9.76}$$

These values will approximate the value of π.

The only remaining issue is to define a convenient starting point of this iterative scheme. We recall that M_n is defined by equation (9.69), and L_n in that equation is the sine of a half-angle α_n for the first polygon in the sequence. We solve equation (9.69) for M_n to get

$$\begin{aligned} M_n &= 2 - 4L_n^2 \\ &= 2 - 4\sin^2\alpha_n. \end{aligned} \tag{9.77}$$

It is best to select the n-polygon in such a way that we can get a simple numerical value for $\sin\alpha_n$. Here, we consider two options:

1. A square, when $n = 4$. Then $\alpha_n = \pi/4$ and $\sin\alpha_n = 1/\sqrt{2}$. We obtain $M_n = 0, M_{2n} = \sqrt{2}$ and so on.

TABLE 9.1
Expressions and numerical values for the perimeter of polygons that are inscribed in a circle.

n	M_n	Expression for P_n	Value
6	1	$3\sqrt{2-M_6}$	3
12	$\sqrt{2+1}$	$6\sqrt{2-M_{12}}$	3.10582854123
24	$\sqrt{2+\sqrt{2+1}}$	$12\sqrt{2-M_{24}}$	3.13262861328
48	$\sqrt{2+\sqrt{2+\sqrt{2+1}}}$	$24\sqrt{2-M_{48}}$	3.1393502030469
96	$\sqrt{2+\sqrt{2+\sqrt{2+\sqrt{2+1}}}}$	$48\sqrt{2-M_{96}}$	3.1410319508905

2. A hexagon, when $n = 6$. Then $\alpha_n = \pi/6$ and $\sin \alpha_n = 1/2$. We obtain $M_n = 1, M_{2n} = \sqrt{2+1}$ and so on.

Table 9.1 gives explicit expressions for the first few terms of the sequence that starts with a hexagon and their numerical values. We see from the last column in the table that values do approach π!

Sometimes we start from exploring mathematical formulas and then arrive at a result that has a value in its own right. This section provides an example of such an exploration. Now, we can state the following elegant result:

Theorem 9.16 *The following sequences produce π:*

$$\lim_{m\to\infty} 6\cdot 2^{m-2}\cdot\sqrt{2-\underbrace{\sqrt{2+\sqrt{2+\sqrt{2+\ldots+\sqrt{2+\sqrt{1}}}}}}_{m}}=\pi,$$
$$\lim_{m\to\infty} 2^{m+1}\cdot\sqrt{2-\underbrace{\sqrt{2+\sqrt{2+\sqrt{2+\ldots+\sqrt{2}}}}}_{m}}=\pi. \tag{9.78}$$

where the number of nested radicals in the expression $\sqrt{2+\ldots}$ is given by m.

The proof of this theorem follows from the derivation above. Compare the two equations in (9.78). The left-hand sides have subtle differences, but both produce π.

Equations (9.78) are examples of a $0\cdot\infty$ indeterminate form. Indeed, from equation (9.65) and the preceding derivation in section 9.10 we see that the value of $\sqrt{2+\ldots\sqrt{2}}$ becomes ever closer to 2 as the number of radicals increases. This means that the value of $\sqrt{2-\sqrt{2+\ldots\sqrt{2}}}$ becomes ever smaller. In equations (9.78), the small values of the expression with radicals is perfectly balanced by the large values of the powers of 2, magically producing π in the product. This is particularly curious because π does not appear anywhere in the right-hand side.

9.12 The Limit for the Base of Natural Logarithms

In section 6.5.1 we introduced a limit for the base of natural logarithms without a proof. In this section we prove that such a limit exists.

Theorem 9.17 *The sequence $e_1, e_2, e_3, \ldots$, where*

$$e_k = \left(1 + \frac{1}{k}\right)^k, \tag{9.79}$$

is convergent.

Proof

We first prove that the sequence of numbers (9.79) is monotonically increasing, and then we prove that it is bounded. Then the limit must exist according to theorem 9.9.

We use the binomial expansion to transform the expression in the right-hand side of equation (9.79). There we use the explicit expressions for binomial coefficients as given by equation (5.20).

$$\begin{aligned}\left(1 + \frac{1}{k}\right)^k &= 1 + k\frac{1}{k} + \frac{k(k-1)}{1 \cdot 2}\frac{1}{k^2} + \frac{k(k-1)(k-2)}{1 \cdot 2 \cdot 3}\frac{1}{k^3} + \ldots \\ &\frac{k(k-1)\ldots(k-(k-1))}{1 \cdot 2 \cdot 3 \cdot \ldots \cdot k}\frac{1}{k^k}.\end{aligned} \tag{9.80}$$

We divide each factor in the numerators by a k in the denominator, such as

$$\begin{aligned}\frac{k(k-1)}{k^2} &= \left(1 - \frac{1}{k}\right), \\ \frac{k(k-1)(k-2)}{k^3} &= \left(1 - \frac{1}{k}\right)\left(1 - \frac{2}{k}\right),\end{aligned} \tag{9.81}$$

and so on. This produces the following expression:

$$\begin{aligned}\left(1 + \frac{1}{k}\right)^k &= 1 + 1 + \frac{1}{2}\left(1 - \frac{1}{k}\right) + \frac{1}{2 \cdot 3}\left(1 - \frac{1}{k}\right)\left(1 - \frac{2}{k}\right) + \ldots \\ &\frac{1}{2 \cdot 3 \cdot \ldots \cdot k}\left(1 - \frac{1}{k}\right)\left(1 - \frac{2}{k}\right)\ldots\left(1 - \frac{k-1}{k}\right).\end{aligned} \tag{9.82}$$

If the value of k is increased, the right-hand side of this equation is changed in two ways:

1. *Each factor in the form $(1 - 1/k), (1 - 2/k)$ and so on gets larger.*[3]

[3] A proof of this statement is the subject of exercise 13 for this chapter.

2. *The number of terms in the sum is increased. Since each term is positive, this increases the sum.*

Therefore, as the value of k is increased, the value of $(1 + 1/k)^k$ grows monotonically. Hence, sequence (9.79) is monotonically increasing.

Now, we proceed to proving that this sequence is bounded. Let's see what happens if in each additive term in equation (9.82) we replace all factors in the denominators by 2, and replace all factors in the parentheses (such as $(1 - 1/k)$ or $(1 - 2/k)$) by 1. For example, we replace

$$\frac{1}{2 \cdot 3}\left(1 - \frac{1}{k}\right)\left(1 - \frac{2}{k}\right) \tag{9.83}$$

with

$$\frac{1}{2 \cdot 2} 1 \cdot 1. \tag{9.84}$$

The denominator here will decrease and the numerator will increase. The net result will be a larger value in the right-hand side:

$$\frac{1}{2 \cdot 3}\left(1 - \frac{1}{k}\right)\left(1 - \frac{2}{k}\right) < \frac{1}{2 \cdot 2}. \tag{9.85}$$

Applying this replacement to all terms in the right-hand side of (9.82) produces the following:

$$\begin{aligned}\left(1 + \frac{1}{k}\right)^k &= 2 + \frac{1}{2}\left(1 - \frac{1}{k}\right) + \frac{1}{2 \cdot 3}\left(1 - \frac{1}{k}\right)\left(1 - \frac{2}{k}\right) + \ldots \\ &\frac{1}{2 \cdot 3 \cdot \ldots \cdot k}\left(1 - \frac{1}{k}\right)\left(1 - \frac{2}{k}\right)\ldots\left(1 - \frac{k-1}{k}\right) \\ &< 2 + \frac{1}{2} + \frac{1}{2^2} + \frac{1}{2^3} + \ldots + \frac{1}{2^{k-1}}.\end{aligned} \tag{9.86}$$

In the right-hand side, we get a 2 plus the familiar geometric sequence, which is bounded (see p. 214):

$$2 + \frac{1}{2} + \frac{1}{2^2} + \frac{1}{2^3} + \ldots + \frac{1}{2^{k-1}} < 3. \tag{9.87}$$

Therefore, the sequence $(1 + 1/k)^k$ is bounded as well:

$$\left(1 + \frac{1}{k}\right)^k < 3. \tag{9.88}$$

We see that sequence $\left(1 + \frac{1}{k}\right)^k$ is monotonically increasing and bounded, therefore it has a limit.

This limit is called the base of natural logarithms and denoted as e.

9.13 Partial Sums and Infinite Series

We are already familiar with sums of finite arithmetic and geometric sequences (see sections 9.1 and 9.2). Using the theory of limits, the idea of summing elements of a sequence can be extended to infinite sequences. Such an infinite sum is called a series. For any series we define a sequence of *partial sums* of N elements, where N spans all natural numbers. For example, consider a geometric series:

$$S_g = 1 + q + q^2 + q^3 + \ldots \tag{9.89}$$

This equation is understood as a limit of the sequence that is formed by partial sums:

$$\begin{aligned} S_g(1) &= 1, \\ S_g(2) &= 1 + q, \\ S_g(3) &= 1 + q + q^2, \\ S_g(4) &= 1 + q + q^2 + q^3, \\ &\ldots \end{aligned} \tag{9.90}$$

If the sequence $\{S_g(1), S_g(2), S_g(3), S_g(4), \ldots\}$ converges, we say that series (9.89) converges.

Theorem 9.18 *For $|q| < 1$, the geometric series converges to*

$$S_g = 1 + q + q^2 + q^3 + \ldots = \frac{1}{1-q}. \tag{9.91}$$

Proof

We recall the formula for the sum of first N elements (section 9.2):

$$\begin{aligned} S_g(N) &= 1 + q + q^2 + q^3 + \ldots + q^N \\ &= \frac{1 - q^{N+1}}{1 - q}. \end{aligned} \tag{9.92}$$

In section 9.5 we proved that sequence $\{1, q, g^2, \ldots\}$ (without the sum) converges to 0 if $|q| < 1$. Therefore, for $N \to \infty$ we have $q^N \to 0$. Separately, in section 9.6 we proved that if a sequence converges, then a sum or a product of this sequence and a constant also converges. Therefore, we conclude that the sum of a geometric series converges to

$$\begin{aligned} \lim_{N\to\infty} S_g(N) &= \lim_{N\to\infty} (1 + q + q^2 + q^3 + \ldots + q^N) \\ &= \lim_{N\to\infty} \frac{1 - q^{N+1}}{1 - q} \\ &= \frac{1}{1-q} \end{aligned} \tag{9.93}$$

for $|q| < 1$.

Again, here the infinite summation should be understood as a limit of the sequence that is formed by partial sums, and not in any other way.

As we will see in section 9.16, we need to be particularly rigorous when working with infinite series. For example, sometimes you may see the following derivation of the formula for the sum of geometric series. We define

$$S_g = \sum_{n=0}^{\infty} q^n = 1 + q + q^2 + q^3 + \dots \tag{9.94}$$

We multiply this equation by q to get

$$qS_g = q + q^2 + q^3 + q^4 + \dots \tag{9.95}$$

Comparing equations (9.94) and (9.95), we see that

$$qS_g = S_g - 1. \tag{9.96}$$

From this we do get the final result (9.91). However, this derivation simply assumes that S_g exists, which is true only for $|q| < 1$. This important condition does not appear anywhere in equations (9.94) through (9.96), creating the impression that formula (9.91) is valid for all $q \neq 1$—that is, when the denominator is not equal to zero. A rigorous derivation does not make any assumptions about the existence of S_g, and the condition $|q| < 1$ pops up there as a necessary one.

Using formula (9.91), we can compute the sum of another series, where the signs of the terms alternate:

$$S'_g = 1 - x + x^2 - x^3 + \dots \tag{9.97}$$

To do that, we just need to introduce $q = -x$. Then series (9.97) takes the familiar form and produces a geometric series (9.91):

$$\begin{aligned} S'_g &= 1 - x + x^2 - x^3 + \dots \\ &= 1 + q + q^2 + q^3 + \dots \\ &= \frac{1}{1-q} \\ &= \frac{1}{1+x} \end{aligned} \tag{9.98}$$

for $|x| < 1$. Another example of summing a series is the subject of the following theorem.

Theorem 9.19 *For* $|x| < 1$, *we have*

$$1 - 2x + 3x^2 - 4x^3 + \dots = \frac{1}{(1+x)^2}. \tag{9.99}$$

Proof

Below we assume that series (9.99) converges, and therefore we can proceed to computing its sum.[4] *We denote:*

$$S_x = 1 - 2x + 3x^2 - 4x^3 + ... \tag{9.100}$$

This is a tough series to crack, and the computation of its sum requires knowledge of calculus or a lucky guess to compute $S_x - S'_g$, where S'_g is defined by equation (9.97):

$$S_x - S'_g = -x + 2x^2 - 3x^3 + ... \tag{9.101}$$

In the right-hand side, we factor out $-x$:

$$S_x - S'_g = -x(1 - 2x + 3x^2 + ...), \tag{9.102}$$

which yields

$$S_x - S'_g = -xS_x. \tag{9.103}$$

We get $(1+x)S_x = S'_g$, and then by using equation (9.98) we get

$$S_x = \frac{1}{(1+x)^2}. \tag{9.104}$$

We will use this result in section 9.17.

9.14 The Harmonic Series

We must not assume that a series always converges if its elements become ever smaller. Consider, for example, the so-called *harmonic series*:

$$S_h = 1 + \frac{1}{2} + \frac{1}{3} + \frac{1}{4} + \frac{1}{5} + \dots \tag{9.105}$$

Every next element produces an ever smaller increase to the partial sum, but this series diverges.

Theorem 9.20 *The harmonic series (9.105) does not converge.*

[4] A proof of convergence for this series is beyond the scope of this book, but its idea is that we can select such a converging geometric series that bounds all elements of series (9.99) by absolute value. That will ensure convergence of series (9.99).

Proof

The proof is by contradiction: we assume that the harmonic series converges. We combine the elements as follows: the first two elements stand alone, then we combine the next two elements, then the next four, and so on:[5]

$$\begin{aligned} S_h = 1 &+ \frac{1}{2} \\ &+ \left(\frac{1}{3} + \frac{1}{4}\right) \\ &+ \left(\frac{1}{5} + \frac{1}{6} + \frac{1}{7} + \frac{1}{8}\right) \\ &+ \ldots \end{aligned} \tag{9.106}$$

In the sum $\left(\frac{1}{3} + \frac{1}{4}\right)$ *both terms are greater or equal that* $\frac{1}{4}$*; therefore, this sum is greater than* $\frac{1}{2}$*. In the sum* $\left(\frac{1}{5} + \frac{1}{6} + \frac{1}{7} + \frac{1}{8}\right)$ *all four terms are greater or equal that* $\frac{1}{8}$*; therefore, this sum is greater than* $\frac{1}{2}$*. This process can be continued to produce an infinite number of terms, each greater than* $\frac{1}{2}$*. Therefore, partial sums grow indefinitely, which contradicts our assumption.*

> The first proof of divergence of this series belongs to Oresme, whom we met when we discussed functions (section 4.1).

9.15 The Harmonic Sequence and Prime Numbers

We are already familiar with sequences that are formed by sums or nested radicals. Here, we introduce a sequence that is formed by products with increasing number of factors. Consider a sequence of numbers $a_1, a_2, \ldots$ Just as we formed partial sums in section 9.13, we can form partial products: the first term is equal to a_1, the second to $a_1 \cdot a_2$, and so on. For such products there is a convenient notation that is analogous to the sigma notation for sums:

$$\prod_{n=1}^{N} a_n = a_1 \cdot a_2 \cdot a_3 \cdot \ldots \cdot a_N. \tag{9.107}$$

Leonhard Euler came up with an interesting proof of the fact that there are infinite number of primes (theorem 2.2). The proof is by contradiction,

[5]As we shall learn in section 9.16, associativity applies to convergent series, but does not apply to divergent ones.

as are the two proofs we saw in section 2.1. We assume that there is a finite number of primes and compute the following product:

$$\prod_{n=1}^{N} \frac{1}{1-\frac{1}{p_n}} = \frac{1}{1-\frac{1}{p_1}} \cdot \frac{1}{1-\frac{1}{p_2}} \cdot \ldots \cdot \frac{1}{1-\frac{1}{p_N}}. \tag{9.108}$$

where $p_1, \ldots, p_N$ are the primes. Note that since this product contains a finite number of factors, the result is finite (later this statement will contradict our conclusion, thus proving the theorem).

Euler expanded each factor in this product into a geometric series. For example,

$$\frac{1}{1-\frac{1}{p_n}} = 1 + \frac{1}{p_n} + \frac{1}{p_n^2} + \frac{1}{p_n^3} + \ldots \tag{9.109}$$

If we substitute all these geometric series into the product that is given by equation (9.108), we will get a rather unruly expression:

$$\prod_{n=1}^{N} \frac{1}{1-\frac{1}{p_n}} = \prod_{n=1}^{N} \left(1 + \sum_{k=1}^{\infty} \frac{1}{p_n^k}\right). \tag{9.110}$$

Let's now think what happens if we distribute all factors in the right-hand side of this equation. Since it is difficult to visualize what happens for this product of multiple infinite series, let's look at a simple case first: we distribute just three pairs of parentheses, with just two or three terms in each of them:

$$\begin{aligned} &\left(1 + \frac{1}{2} + \frac{1}{2^2}\right) \cdot \left(1 + \frac{1}{3}\right) \cdot \left(1 + \frac{1}{5}\right) \\ &= 1 + \frac{1}{2} + \frac{1}{2^2} + \frac{1}{3} + \frac{1}{2 \cdot 3} + \frac{1}{2^2 \cdot 3} \\ &+ \frac{1}{5} + \frac{1}{2 \cdot 5} + \frac{1}{2^2 \cdot 5} + \frac{1}{3 \cdot 5} + \frac{1}{2 \cdot 3 \cdot 5} + \frac{1}{2^2 \cdot 3 \cdot 5}. \end{aligned} \tag{9.111}$$

In the denominators, we get products of various combinations of prime numbers 2, 3, and 5. If we expand the full product in the right-hand side of equation (9.110), with all the prime numbers and all the terms in the infinite series, we will get a sum of infinite number of terms. The denominator of each term will also contain a product of some combination of primes.

So far it looks like we just have made the equations more complicated. The genius of Euler was to find an amazing simplification of this infinite sum. If we think about it, we will see that denominators there contain all possible products of prime numbers, with these prime numbers raised to all possible powers. This is true because all powers and all primes are present in the parentheses in the right-hand side of equation (9.110). Moreover, no matter what combination of primes we select, it will match one and only one of the denominators in the infinite sum. In addition to the terms listed in equation (9.111), there will be terms, whose denominators contain all other primes in all kinds of powers, such as $1/(2 \cdot 7^2)$ or $1/(17 \cdot 41^2 \cdot 797^5)$.

The next step is to invoke the Fundamental Theorem of Arithmetic (theorem 2.1) that states that every natural number can be uniquely represented as a product of primes. Therefore, for any natural number there is one and only one term in the sum that will have that number in the denominator. For example, for number 98 there will be a term in the sum that contains $1/(2 \cdot 7 \cdot 7)$. That term can be traced back to the expansion of the right-hand side of equation (9.110), where we would have to pull 1 from every pair of parentheses, except parentheses for primes 2 and 7, and from those we pull the term $1/2$ for the prime 2 and the term $1/7^2$ for the prime 7.

We see that the expansion of the right-hand side of (9.110) simply contains reciprocals of all natural numbers, and each natural number is present there only once. If we sort all the terms in the descending order, we will get the following elegant result:

$$\prod_{n=1}^{N} \frac{1}{1 - \frac{1}{p_n}} = 1 + \frac{1}{2} + \frac{1}{3} + \dots \tag{9.112}$$

This is a familiar harmonic series!

We know that the harmonic series diverges, and in the beginning of this section we concluded that the right-hand side in equation (9.108) computes to a finite value, which is a contradiction. This proves that there are infinite number of primes.

This proof looks like an overkill for a theorem that we have already proved, but we consider it here because it contains two amazing features. First, we have the beautiful collapse of a very complicated expression into a harmonic series. Second, this proof somehow links product (9.108), whose factors are enumerated by primes, and the harmonic series, whose elements are enumerated by natural numbers. This looks like a way of bringing a bit of order to the irregularly spaced primes. Indeed, the line of thinking in this proof has become a stepping stone in proving other important theorems about prime numbers.

Leonhard Euler worked in many different areas of mathematics, physics, and astronomy, producing close to 900 scientific works: papers, books, and important letters. However, his prolific legacy may have been surpassed by Paul Erdős, who published about 1,500 papers.

9.16 Intuition May Fail Us for Infinite Series

Using mathematical induction, we can extend associativity and commutativity to any finite number of terms in a sum. This prompts a question: are these properties valid for infinite series? The answer is no. Indeed, consider a series

$$S_{\pm 1} = 1 - 1 + 1 - 1 + 1 - 1 + \ldots \tag{9.113}$$

Let's combine terms in this series in two different ways:

$$\begin{aligned} S_{\pm 1} &= (1-1) + (1-1) + (1-1) + \ldots = 0, \\ S'_{\pm 1} &= 1 + (-1+1) + (-1+1) + \ldots = 1. \end{aligned} \tag{9.114}$$

We get a contradiction, which means that associativity cannot be applied. This is an example of the need for rigor in mathematics. While we can prove the associativity property for finite sums, an extension to infinite series cannot be proved, and familiar intuitive rules do not apply. (However, associativity does apply to convergent series.)

Similarly, we cannot always apply commutativity to infinite series. This is formalized by the famous *Riemann series theorem.* To introduce this theorem, we need to define a *conditionally convergent series.* Consider a convergent series, where the terms have different signs:

$$S = a_1 + a_2 + a_3 + \ldots, \tag{9.115}$$

that is, some a_n are positive and some are negative. Next, construct a new series using the absolute values of the same terms:

$$\tilde{S} = |a_1| + |a_2| + |a_3| + \ldots. \tag{9.116}$$

Here all terms are nonnegative. Sometimes series S is convergent, but $\tilde{S}$ is not; then series S is called conditionally convergent. For example, this happens if we alternate signs in the harmonic series. We already know that the harmonic series diverges (theorem 9.20). However, if the signs of the terms in a harmonic series alternate, it does converge:

$$1 - \frac{1}{2} + \frac{1}{3} - \frac{1}{4} + \ldots \to \ln 2. \tag{9.117}$$

Therefore, the alternating harmonic series is conditionally convergent.

The Riemann series theorem states that it is possible to rearrange the terms of a conditionally convergent series in such a way, that the series will diverge or it will converge to any number we choose! For example, it is possible to shuffle the order of summation of the alternating harmonic series for the series to diverge, or converge to -0.1, π, 10^{10}, or any other number!

Another great example of a counterintuitive mathematical statement that includes an infinite number of objects is the Banach-Tarski paradox. It says that we can take a regular three-dimensional ball, cut it into infinite number of pieces, rotate and shift these pieces around, and then reassemble them in such a way, that we will get *two* balls, each having the same volume as the original ball.

This looks weird because of two major assumptions that work for us every day when we cut produce for a salad and that contribute to our intuitive thinking:

1. If we cut a 3D shape into pieces, the sum of volumes of the pieces is equal to the volume of the original shape. Same is true when we put 3D pieces together: the total volume is preserved.
2. Rotations and shifts do not change the volume of any geometrical objects.

Yet, in Banach-Tarski paradox we somehow get a second, equally large ball out of nothing! There is a formal proof of this fact, which is beyond the scope of this book. A careful look into the proof reveals that the total volume is preserved only if we have a finite number of pieces. For some (but not all) sets of objects containing infinite number of pieces, the volume is not defined at all. Therefore, there is no expectation that the volume will be preserved after we reassemble these pieces.

9.17 Sometimes Neglecting Rigor Is a Good Thing

In this section we look at a totally nonrigorous derivation, which produces a nonsensical result. Surprisingly, this result turns out to be less absurd than it seems at the first glance. This section should not be used as an excuse to neglect rigor in mathematics, but it illustrates a situation when a flimsy derivation produces a new insight into a problem.

Below we suspend all guards about convergence and assume that all series that are considered in this derivation converge (they do not). We compute the sum of all natural numbers:

$$S_{\{N\}} = 1 + 2 + 3 + 4 + \dots \tag{9.118}$$

Next, we compute $4S_{\{N\}}$:

$$4S_{\{N\}} = 4 + 8 + 12 + 16 + \dots \tag{9.119}$$

Let us subtract the last two equations and group the terms in a particular way:

$$S_{\{N\}} - 4S_{\{N\}} = 1 + (2 - 4) + 3 + (4 - 8) + \dots \tag{9.120}$$

This produces

$$-3S_{\{N\}} = 1 - 2 + 3 - 4 + \dots \tag{9.121}$$

This happens to be series S_x as defined by equation (9.100), where we set $x = 1$. We then use $x = 1$ in equation (9.104) to get

$$-3S_{\{N\}} = \frac{1}{(1+1)^2} = \frac{1}{4}. \tag{9.122}$$

This produces an obviously absurd result:

$$S_{\{N\}} = -\frac{1}{12}. \tag{9.123}$$

(Note that series (9.118) also looks like the left-hand side of equation (9.99), where $x = -1$, but we cannot use the right-hand side of that equation for computing this sum because we would get a zero in the denominator.)

In the course of this "derivation," we made several mistakes, all of which were connected to the assumption of convergence. However, this absurd derivation has an unexpected positive side effect: the final result, while erroneous, happens to provide a peek into properties of the so-called Riemann zeta function, which we will introduce in exercise 23 below. Our neglect of mathematical rigor has produced something of value!

Of course, any decent computation of the zeta function cannot be based on a derivation like the one above; rather, it must be made using different (and much more complicated) methods.

Exercises

1. Prove equation (9.2) from the definition of arithmetic sequence (9.1).
2. Prove equation (9.10) by multiplying it by the denominator and expanding the product in the left-hand side. Do you detect similarities with one of the proofs presented in section 9.2?
3. Prove that $a^n - b^n = (a-b)\sum a^j b^{n-1-j}$ by using a formula for the sum of a geometric sequence.
4. Assume that series

$$S_0 = 1 - \frac{1}{2^2} - \frac{1}{4^2} - \frac{1}{6^2} - \frac{1}{8^2} \cdots \tag{9.124}$$

and

$$S_1 = 1 - \frac{1}{2} + \frac{1}{3} - \frac{1}{4} + \frac{1}{5} \cdots \tag{9.125}$$

are convergent. Prove that $S_0 < S_1$. (In fact, $S_0 = 1 - \pi^2/24 \approx 0.588766$, and S_1 is given by equation (9.117): $S_1 = \ln 2 \approx 0.693147$.)

5. Consider a sum $\sum_{n=1}^{N}(-1)^{n-1}n = 1-2+3-4+5-...+(-1)^{N-1}N$. Compute the value of this sum for several values of N and detect a pattern. Prove that this pattern holds for any value of N using induction.
6. Consider two arithmetic sequences $\{S_n\}$ and $\{R_n\}$ of equal lengths, where $S_n = S_1+(n-1)D$ and $R_n = R_1+(n-1)C$. Define sequence $\{T_n\}$ that is a pairwise sum of these two arithmetic sequences:

$$T_n = S_n + R_n = (S_1 + R_1) + (n-1)(D + C). \tag{9.126}$$

   Obtain explicit expressions for the sums $\tilde{S} = \sum S_n, \tilde{R} = \sum R_n$ and $\tilde{T} = \sum T_n$ of all three sequences using theorem 9.1. Then prove that these expressions satisfy $\tilde{S} + \tilde{R} = \tilde{T}$.
7. Prove the following formulas:
   (a) $\sum_{k=1}^{n}(-1)^k k^2 = (-1)^n n(n+1)/2$.
   (b) $\sum_{k=1}^{n} k^2 = \frac{1}{6}n(n+1)(2n+1)$.
   (c) $\sum_{k=1}^{n} k^3 = \frac{1}{4}n^2(n+1)^2$.
8. Consider geometric sequence 1, 2, 4, 8, ... Prove that there is an integer representation of any natural number N using a sum of elements of this sequence that has nonrepeating entries. (Hint: for any number N consider the largest element a_k of the geometric sequence that does not exceed N, then compute $N - a_k$ and continue this process.)
9. Sometimes you may come across a puzzle, where you need to identify the next number in a sequence of several numbers. For such puzzles, you first need to guess the underlying rule for forming that sequence. For example, you are given $2, 5, 26, 677, 458330, ...$ (The answer supposedly is: $5 = 2^2+1; 26 = 5^2+1; 677 = 26^2+1; 458330 = 677^2+1$.)

   Why numbers $2, 5, 26, 677, 458330, ...$ cannot be considered a mathematical sequence in the rigorous meaning of this term?
10. Prove that the sum of M convergent sequences converges to the sum of their corresponding limits.
11. Provide a direct proof that radicals are monotonically increasing: consider $\frac{x-y}{\sqrt[n]{x}-\sqrt[n]{y}}$ and factorize the numerator. (Assume $x > y \geq 0$.)
12. Can a sequence be both monotonically increasing and monotonically decreasing?
13. On p. 233 it is stated that for $n, k > 0$, expression $(1-n/k)$ increases if k increases. Prove this statement.
14. Consider the infinite sequence that is formed by using the digits of π on p. 223. Prove that it contains at least one convergent subsequence without using the Bolzano-Weierstrass theorem. (Hint: use a proof by contradiction.)

15. (a) Provide such examples of pairs of sequences that each converges to zero and that their ratio converges to zero, to a finite value, or diverges.
    (b) Give an example of two diverging sequences, such that their ratio converges to zero, to a finite value, or diverges.
16. Derive a formula for the sum of odd numbers $\sum_{n=1}^{N}(2n-1)$ using two methods:
    (a) Use the identity $n^2-(n-1)^2=2n-1$.
    (b) Use the formula for the sum of arithmetic sequence.
17. It is known that
$$\sin\frac{\pi}{10}=\frac{\sqrt{5}-1}{4}. \quad (9.127)$$
Using this value, derive an approximation for π that starts with a pentagon and then successively doubles the number of sides.
18. Consider a convergent sequence $\{c_n\}$ that converges to some number C, and a polynomial function $f(x)=a_nx^N+a_{N-1}x^{N-1}+...+a_1x+a_0$. Prove that sequence $\{f(c_n)\}$ converges to $f(C)$.
19. Prove theorems 9.5 and 9.6.
20. Prove equation (9.3) for odd values of N.
21. Follow the line of arguments for case 1a (for $|q|<1$) in the proof of theorem 9.3, and modify it for case 3b (that is, for $|q|>1$).
22. Equation (9.45) has a flaw, but this flaw can be fixed without jeopardizing the conclusion that sequence $\sin(1),\sin(2),\sin(3),\sin(4),...$ has more than one convergent subsequence. Find that flaw and modify the proof.
23. The Riemann zeta function is defined as
$$\zeta(s)=\sum_{n=1}^{\infty}\frac{1}{n^s}. \quad (9.128)$$
The series in the right-hand side converges for $|s|>1$. Use the derivation in section 9.15 as a template to show that
$$\zeta(s)=\prod_{k=1}^{\infty}\frac{1}{1-\frac{1}{p_k^s}}, \quad (9.129)$$
where p_k is the k-th prime number. Which properties of exponentiation you had to use in this derivation?
24. Prove that sequence $0,\frac{1}{2},\frac{2}{3},\frac{3}{4},\frac{4}{5},...$ converges to 1.

25. Prove that if a sum $a_n + b_n$ of two sequences a_n and b_n is convergent, then a_n and b_n are either both convergent or both divergent. Why a similar statement is not true for a product?

26. Prove that for any natural N and for any positive q, such that $q \neq 1$, it is the case that
$$N \leq \frac{1 - q^{2N}}{q^{N-1}(1 - q^2)}. \tag{9.130}$$
(Hint: use the Cauchy-Schwarz inequality (3.38) and set $x_n = q^{n-1}, y_n = q^{N-n}$ there.) Can you obtain the same result from the solution of exercise 29 in chapter 3?

27. Any complex number q with a unit modulus ($|q| = 1$) can be represented as $q = e^{i\theta}$. Consider exercise 29 in chapter 3. Prove that if q there is a complex number, $|q| = 1$, then the sign of the inequality in that exercise flips:
$$q^{-n} + q^{-n+2} + \ldots + q^{n-2} + q^n \leq n + 1. \tag{9.131}$$
(Note that the sum in the left-hand side is real.)

28. Compute the sum in inequality (9.131) for complex q, where q is such that $|q| = 1$. In the result, represent q as $q = e^{i\phi}$ and use equations (7.66) to prove theorem 7.16.

29. Consider geometric series (9.91) when $|q| < 1$. Group pairs of terms $(q + q^2)$, $(q^3 + q^4)$ and so forth, and sum the series
$$\begin{aligned} &1 + (q + q^2) + (q^3 + q^4) + \ldots = \\ &1 + q(1 + q) + q^3(1 + q) + \ldots \end{aligned} \tag{9.132}$$
that is produced by this grouping. Show that the result is the same as for the original geometric series. Why this cannot serve as a proof that associativity works for the geometric series?

10

The Fibonacci Sequence

The Fibonacci sequence is defined recursively. The first two elements are equal to 1. Each following element is equal to the sum of the two previous ones:

$$F_n = F_{n-1} + F_{n-2}. \tag{10.1}$$

The first few elements of this sequence are

$$F = \{1, 1, 2, 3, 5, 8, 13, \ldots\}. \tag{10.2}$$

They are called Fibonacci numbers. This sequence grows indefinitely and therefore does not converge. Let's explore the wealth of features buried in this seemingly unremarkable sequence. It possesses a number of interesting properties, and proofs of some of these properties are quite elegant.

10.1 Cassini's Identity

This concise identity links products of Fibonacci numbers.

Theorem 10.1 *Fibonacci numbers satisfy the following identity:*

$$F_{n-1}F_{n+1} - F_n^2 = (-1)^n. \tag{10.3}$$

> This identity bears the name of the Italian mathematician and astronomer Giovanni Domenico Cassini, who also made important discoveries about the planet Saturn. NASA named its 1997 mission to Saturn after Cassini.

Proof

The proof is by induction.

1. *For* $n = 2$ *we get*

$$F_1F_3 - F_2^2 = (-1)^2. \tag{10.4}$$

We substitute $F_1 = 1, F_2 = 1, F_3 = 2$ *and verify this case.*

DOI: 10.1201/9781003456766-10

2. *We assume that Cassini's identity holds for n and rewrite equation (10.3) for n + 1:*

$$F_n F_{n+2} - F_{n+1}^2 = (-1)^{n+1}. \tag{10.5}$$

The second term in the left-hand side is a product of two identical factors:

$$F_n F_{n+2} - F_{n+1} F_{n+1} = (-1)^{n+1}. \tag{10.6}$$

We substitute values of F_{n+2} *and of one of the factors in the second term by using equation (10.1). We leave the second factor* F_{n+1} *in the second term intact:*

$$F_n(F_{n+1} + F_n) - F_{n+1}(F_n + F_{n-1}) = (-1)^{n+1}. \tag{10.7}$$

We expand the parentheses and collect the terms. This yields:

$$F_n^2 - F_{n+1} F_{n-1} = (-1)^{n+1}, \tag{10.8}$$

which is equivalent to Cassini's identity for n. This proves the theorem.

10.2 The Golden Ratio

Let's define a sequence of ratios of successive Fibonacci numbers:

$$\left\{ \frac{F_2}{F_1}, \frac{F_3}{F_2}, \frac{F_4}{F_3}, \ldots \right\}. \tag{10.9}$$

Let's assume for now that this sequence converges (a proof of convergence will be given later in this chapter). Using the convergence assumption, we can find the limit of this sequence. We denote the limit as ϕ:

$$\lim_{n\to\infty} \frac{F_n}{F_{n-1}} = \phi. \tag{10.10}$$

This means that for large n we have

$$\begin{aligned} F_n &\approx \phi F_{n-1} \\ &\approx \phi^2 F_{n-2}. \end{aligned} \tag{10.11}$$

Substitution of (10.11) in (10.1) produces

$$\phi^2 F_{n-2} \approx \phi F_{n-2} + F_{n-2} \tag{10.12}$$

or

$$\phi^2 - \phi - 1 \approx 0. \tag{10.13}$$

This equation becomes ever more accurate for larger values of n. In the limit $n \to \infty$, we get

$$\phi^2 - \phi - 1 = 0. \tag{10.14}$$

This is a quadratic equation for ϕ, and it has two solutions. One of these solutions is positive and another is negative. Since we know that all elements of the Fibonacci series are positive, we must select the positive solution for ϕ, which is given by theorem 5.5:

$$\phi = \frac{1+\sqrt{5}}{2} \approx 1.618034. \tag{10.15}$$

Constant ϕ is called the *golden ratio*. The second solution is

$$\psi = \frac{1-\sqrt{5}}{2} \approx -0.618034. \tag{10.16}$$

From Vieta's theorem (p. 125) we know that

$$\begin{aligned} \phi + \psi &= 1, \\ \phi\psi &= -1. \end{aligned} \tag{10.17}$$

From equations (10.17) or directly from equation (10.13) we prove the following property for the golden ratio:

Theorem 10.2 *If numbers a and b are such that $a/b = \phi$, then $(a+b)/a = b/(a-b) = \phi$.*

Proof

We consider the value of $(a+b)/a$:

$$\frac{a+b}{a} = 1 + \frac{b}{a}. \tag{10.18}$$

From $a/b = \phi$ and from the second equation in (10.17) we obtain:

$$\frac{a+b}{a} = 1 - \psi. \tag{10.19}$$

Then

$$\frac{a+b}{a} = \phi \tag{10.20}$$

because of the first equation in (10.17). A proof of equation $b/(a-b) = \phi$ is the subject of exercise 13 for this chapter.

> The golden ratio is used in computer optimization algorithms. Such algorithms "try" function values at various points, seeking a minimum. The properties of the golden ratio allow to decrease the number of function computations, making the algorithm faster.

Both ϕ and ψ are irrational numbers. This, of course, follows from theorem 2.13. In addition, here is an interesting proof for the irrationality of ϕ that is based on theorem 10.2.

Theorem 10.3 *The golden ratio ϕ is irrational.*

Proof

We use a proof by contradiction—that is, we assume $\phi = m/n$, where m, n are natural numbers. If $\phi = m/n$ is expressed in the canonical form,[1] then $n/(m-n)$ has the same value: $n/(m-n) = \phi$, but it uses even smaller values for the numerator and denominator, which is a contradiction.

Note a bit of similarity between this proof and one of the proofs for the irrationality of $\sqrt{2}$ (see p. 37).

10.3 The Golden Ratio via Nested Radicals

We already came across number ϕ in section 9.10. There we showed that

$$a_n = \underbrace{\sqrt{1+\sqrt{1+\sqrt{1+\ldots\sqrt{1}}}}}_{n \text{ times}} \tag{10.21}$$

approaches $(1+\sqrt{5})/2$ as $n \to \infty$. Table 10.1 gives first several values of this sequence. These values do approach $\phi \approx 1.618034$. Another way to look at sequence (10.21) is to notice that $a_{n+1} = \sqrt{1+a_n}$. As $n \to \infty$, we get $a_n \to \phi$ and $a_{n+1} \to \phi$. Then equation $a_{n+1} = \sqrt{1+a_n}$ becomes $a = \sqrt{1+a}$, which is equivalent to the definition of the golden ratio (10.14) where ϕ is replaced by a.

[1] See p. 28 for the definition of a rational number being expressed in the canonical form.

TABLE 10.1
Approximation of the golden ratio by nested radicals.

n	**Expression**	**Approximate value**
1	$\sqrt{1}$	1
2	$\sqrt{1+\sqrt{1}}$	1.41421356
3	$\sqrt{1+\sqrt{1+\sqrt{1}}}$	1.55377397403
4	$\sqrt{1+\sqrt{1+\sqrt{1+\sqrt{1}}}}$	1.59805318248
5	$\sqrt{1+\sqrt{1+\sqrt{1+\sqrt{1+\sqrt{1}}}}}$	1.61184775412525
6	$\sqrt{1+\sqrt{1+\sqrt{1+\sqrt{1+\sqrt{1+\sqrt{1}}}}}}$	1.616121206508
7	$\sqrt{1+\sqrt{1+\sqrt{1+\sqrt{1+\sqrt{1+\sqrt{1+\sqrt{1}}}}}}}$	1.61744279852739

10.4 Successive Powers

Here, we explore some properties of constants ϕ, ψ without claiming that ϕ is the limit of the ratios of successive Fibonacci numbers (as we have assumed in equation (10.10)). For the purposes of this section, ϕ and ψ are simply the roots of quadratic equation (10.14). Let's compute successive powers of the golden ratio and simplify each result. From equation (10.14) we get

$$\phi^2 = \phi + 1. \tag{10.22}$$

For the cubic power we substitute ϕ^2 in ϕ^3 to get

$$\begin{aligned}\phi^3 &= \phi \cdot \phi^2 \\ &= \phi(\phi + 1) \\ &= \phi^2 + \phi \\ &= 2\phi + 1,\end{aligned} \tag{10.23}$$

where we used $\phi^2 = \phi + 1$ again. Next, we use the expression for ϕ^3 to compute ϕ^4:

$$\begin{aligned} \phi^4 &= \phi \cdot \phi^3 \\ &= \phi(2\phi + 1) \\ &= 2\phi^2 + \phi \\ &= 3\phi + 2. \end{aligned} \tag{10.24}$$

This way each next power can be expressed in the form $\phi^n = k\phi + m$, where k and m are integers. The following theorem shows how the values of k and m are computed.

Theorem 10.4 *For $n \geq 2$, successive powers of the golden ratio are given by*

$$\phi^n = F_n \phi + F_{n-1}, \tag{10.25}$$

where F_n are the Fibonacci numbers.

Proof
The proof is by induction.

1. *For $n = 2$ the statement of the theorem is true due to equation (10.14).*
2. *We assume that the statement of the theorem is true for n. Then, we get for $n + 1$:*

$$\begin{aligned} \phi^{n+1} &= \phi\phi^n \\ &= \phi\left(F_n\phi + F_{n-1}\right) \\ &= F_n\phi^2 + F_{n-1}\phi \\ &= F_n(\phi + 1) + F_{n-1}\phi \\ &= (F_n + F_{n-1})\phi + F_n \\ &= F_{n+1}\phi + F_n. \end{aligned} \tag{10.26}$$

Curiously, the other root of equation (10.14) follows the same pattern! This is formalized by the following theorem.

Theorem 10.5 *For $n \geq 2$, successive powers of $\psi = (1 - \sqrt{5})/2$ are given by*

$$\psi^n = F_n \psi + F_{n-1}, \tag{10.27}$$

where F_n are the Fibonacci numbers.

Proof
The proof is completely analogous to that of theorem 10.4.

The theorems about successive powers of ϕ and ψ help prove an explicit formula for the Fibonacci numbers:

Theorem 10.6 *Fibonacci numbers are computed as*

$$F_n = \frac{1}{\sqrt{5}}\left(\left(\frac{1+\sqrt{5}}{2}\right)^n - \left(\frac{1-\sqrt{5}}{2}\right)^n\right). \tag{10.28}$$

This is known as Binet's formula.

Proof
We consider the right-hand side of equation (10.28) and recognize the expressions (10.15) and (10.16) for ϕ *and* ψ *in the right-hand side. We use theorems 10.4 and 10.5:*

$$\begin{aligned}&\frac{1}{\sqrt{5}}\left(\left(\frac{1+\sqrt{5}}{2}\right)^n - \left(\frac{1-\sqrt{5}}{2}\right)^n\right) = \\ &\frac{1}{\sqrt{5}}\left(F_n\phi + F_{n-1} - F_n\psi - F_{n-1}\right).\end{aligned} \tag{10.29}$$

The right-hand side simplifies to

$$\frac{1}{\sqrt{5}}\left(F_n\phi + F_{n-1} - F_n\psi - F_{n-1}\right) = \frac{1}{\sqrt{5}}F_n(\phi - \psi). \tag{10.30}$$

We substitute back expressions for ϕ, ψ *to get*

$$\frac{1}{\sqrt{5}}F_n(\phi - \psi) = \frac{1}{\sqrt{5}}F_n\left(\frac{1+\sqrt{5}}{2} - \frac{1-\sqrt{5}}{2}\right). \tag{10.31}$$

The right-hand side of this equation simplifies to F_n*, which proves the theorem.*

Binet's formula has a curious ramification. Consider Fibonacci numbers $F_{10}, F_{100}, F_{1000}, \ldots F_{10^n}, \ldots$ The number of digits in these numbers are, correspondingly, $2, 21, 209, 2090, 20899, \ldots$. These are the digits in successive decimal approximations of $\log_{10}\phi \approx 0.2089876402499787337\ldots$

10.5 A Proof of Convergence

Here, we go back to the ratios of the Fibonacci numbers and prove equation (10.10).

Theorem 10.7 *The ratios of successive Fibonacci numbers converge to the golden ratio:*

$$\lim_{n\to\infty}\frac{F_n}{F_{n-1}} = \phi. \tag{10.32}$$

Proof

We use theorem 10.5 and consider the value of ψ^n there. Since $|\psi| < 1$, this value approaches zero for large n. Therefore,

$$\lim_{n\to\infty} (F_n\psi + F_{n-1}) = 0. \tag{10.33}$$

Then

$$\lim_{n\to\infty} \left(\frac{F_n}{F_{n-1}}\psi + 1\right) = 0. \tag{10.34}$$

From this we obtain

$$\lim_{n\to\infty} \frac{F_n}{F_{n-1}} = -\frac{1}{\psi}. \tag{10.35}$$

Since we have already established that $-1/\psi = \phi$, we get equation (10.32).

10.6 More about Successive Powers

Theorem 10.4 leads to another cool fact. Let's compute powers of ϕ and see what happens. Some results are presented in table 10.2. We see an interesting pattern: these values become close to integers. Can we prove this fact for all n?

To express this pattern mathematically, we should formalize the idea of "closeness" to an integer. For any real number x we define a *rounded* number $\lfloor x \rceil$ as the integer that is closest to x (that is, minimizes $|x - \lfloor x \rceil|$. For example, $\lfloor 1.01 \rceil = 1$. Then $|1.01 - \lfloor 1.01 \rceil| = 0.01$. If a number is close to an integer, the difference between this number and its rounded value is close to zero.

Theorem 10.8 *For large n, the difference between ϕ^n and its rounded value approaches zero:*

$$\lim_{n\to\infty} (\phi^n - \lfloor \phi^n \rceil) = 0. \tag{10.36}$$

TABLE 10.2
Integer powers of the golden ratio constant.

n	ϕ^n (approximate value)
5	11.09016994
10	122.991869381
15	1364.0007331
20	15126.9999339
25	167761.00000596
50	28143753122.99999999996

(Note that values of ψ^n also approach an integer for large values of n. The value of that integer is zero.)

Proof
We start from two familiar equations:

$$\begin{aligned} \phi^n &= F_n\phi + F_{n-1}, \\ \psi^n &= F_n\psi + F_{n-1}. \end{aligned} \tag{10.37}$$

We add these equations to get

$$\phi^n + \psi^n = F_n(\phi + \psi) + 2F_{n-1}. \tag{10.38}$$

Here, we use $\phi + \psi = 1$ and rearrange the terms:

$$\phi^n - (F_n + 2F_{n-1}) = -\psi^n. \tag{10.39}$$

As $n \to \infty$, we have $\psi^n \to 0$, and the difference between ϕ^n and the integer number $F_n + 2F_{n-1}$ becomes ever smaller.

The golden ratio is not the only irrational number that has this property. It is one of the so-called Pisot–Vijayaraghavan numbers. These numbers are irrational, but their n-th powers approach integers for large n.

10.7 Integers as Sums of Fibonacci Numbers

We can represent any natural number N as a sum of Fibonacci numbers. To do that, we use a scheme similar to that in theorem 2.9: we start from the largest Fibonacci number F_k that does not exceed N, then compute $N - F_k$ and continue this process until we get a zero. For example, $50 = 34 + 13 + 3$. This way to represent integers has a curious property.

Theorem 10.9 *There is a representation of any natural number via a sum of Fibonacci numbers such that:*

1. *The terms of the sum do not repeat and*
2. *The sum does not contain adjacent Fibonacci numbers.*

Proof
Consider some integer N. We select the largest Fibonacci number F_n not exceeding N. There are two possibilities:

1. *$F_n = N$. Then the statement of the theorem is true.*

2. *$F_n < N$. Then, we consider a positive integer $M = N - F_n$ and seek the largest Fibonacci number F_m that does not exceed M. As we continue, it is clear that at each next step we deal with a smaller natural number to represent as a sum of Fibonacci numbers. This process may not continue indefinitely and must terminate after a finite number of steps. Even if we never hit some Fibonacci number that is greater than 1, we will invariably end at $F_1 = 1$.*

   Without loss of generality, we consider the first two steps as introduced above: $F_n < N$ and F_m does not exceed $M = N - F_n$. A sum of F_m and F_n would still not exceed N. Mathematically this is expressed as

$$F_m + F_n \leq N. \tag{10.40}$$

   To prove both statements of the theorem, we just need to prove that m is not equal or adjacent to n, which in turn means that $n-m \geq 2$. This proof is by contradiction. We assume that $m \geq n - 1$. Since the Fibonacci sequence is growing, this means that

$$F_m \geq F_{n-1}. \tag{10.41}$$

   From inequalities (10.40) and (10.41) we get

$$F_{n-1} + F_n \leq N. \tag{10.42}$$

   However, $F_{n-1} + F_n = F_{n+1}$, which means that $F_{n+1} \leq N$, and that F_n is not the largest Fibonacci number not exceeding N. This contradiction proves the theorem.

Note that while this theorem is true for the Fibonacci numbers, it is not true for many other sequences, such as powers of 2 or powers of 10. For powers of 10, the terms in the sum may repeat (for example, $20 = 10+10$). For powers of 2, the terms of the sum do not repeat, but the sum may contain neighboring values (for example, $12 = 2^3 + 2^2$). A representation of integers through a sum of Fibonacci numbers is special, and in a sense more economical. The properties given by theorem 10.9 make Fibonacci numbers useful in speeding up computer search algorithms.

10.8 The Partial Sum of Fibonacci Numbers

Theorem 10.10 *The partial sum of Fibonacci numbers is given by*

$$\sum_{n=1}^{N} F_n = F_{N+2} - 1. \tag{10.43}$$

For example,

$$\begin{aligned} F_1 + F_2 &= 1 + 1 = 2 = F_4 - 1, \\ F_1 + F_2 + F_3 &= 1 + 1 + 2 = F_5 - 1, \\ F_1 + F_2 + F_3 + F_4 &= 1 + 1 + 2 + 3 = 7 = F_6 - 1. \end{aligned} \tag{10.44}$$

Proof

The proof is by induction:

1. *For $N = 1$ and $N = 2$ the statement of the theorem holds.*
2. *We assume that it holds for N and consider the sum of $N + 1$ Fibonacci numbers. We separate the first N terms and the $(N+1)$-th term:*

$$\sum_{n=1}^{N+1} F_n = \sum_{n=1}^{N} F_n + F_{N+1}. \tag{10.45}$$

We already know the expression for the sum of the first N terms. We use it in the last equation

$$\sum_{n=1}^{N+1} F_n = F_{N+2} - 1 + F_{N+1}. \tag{10.46}$$

In the right-hand side, we group terms F_{N+2} and F_{N+1} and note that they produce F_{N+3}. This proves the theorem.

10.9 Continued Fractions

Here, we consider a sequence formed by the following iterative procedure:

$$\begin{aligned} a_1 &= 1 \\ a_n &= 1 + \frac{1}{a_{n-1}}. \end{aligned} \tag{10.47}$$

Explicit equations for elements of this sequence form an interesting pattern:

$$\begin{aligned} a_1 &= 1, \\ a_2 &= 1 + \frac{1}{1}, \\ a_3 &= 1 + \frac{1}{1 + \frac{1}{1}}, \\ a_4 &= 1 + \frac{1}{1 + \frac{1}{1 + \frac{1}{1}}}. \end{aligned} \tag{10.48}$$

As n becomes large, we get a large number of levels in the fraction in the right-hand side. Below we prove that this sequence has a limit. For $n \to \infty$, it is customary to write the limit of this sequence in the form of a *continued fraction*:

$$\phi = 1 + \cfrac{1}{1 + \cfrac{1}{1+\cfrac{1}{1+\cfrac{1}{\ddots}}}}. \tag{10.49}$$

For continued fractions we often use a notation that is more concise and avoids dealing with cumbersome multilevel expressions. Consider a continued fraction

$$P = b_1 + \cfrac{1}{b_2 + \cfrac{1}{b_3+\cfrac{1}{b_4+\cfrac{1}{\ddots}}}}. \tag{10.50}$$

We will denote this fraction as

$$P = [b_1; b_2, b_3, b_4, \ldots]. \tag{10.51}$$

In this notation the continued fraction (10.48) is written as

$$\phi = [1; 1, 1, 1, \ldots]. \tag{10.52}$$

The first few numerical values of this continued fraction are as follows:

$$\begin{aligned} a_1 &= 1, \\ a_2 &= 1 + \frac{1}{1} = 2, \\ a_3 &= 1 + \frac{1}{2} = \frac{3}{2}, \\ a_4 &= 1 + \frac{1}{\frac{3}{2}} = \frac{5}{3}, \\ a_5 &= 1 + \frac{1}{\frac{5}{3}} = \frac{8}{5}. \end{aligned} \tag{10.53}$$

We see that they oscillate and cluster at some number between 1.5 and 2. More precisely, we prove the following theorem:

Theorem 10.11 *Sequence (10.47) converges to the golden ratio* ϕ.

Below are two proofs of this theorem. The first proof is based on the observation that values of sequence elements in equations (10.53) look like successive ratios of Fibonacci numbers. We prove this fact by induction:

Proof

1. *We do get* $a_1 = F_2/F_1$ *and* $a_2 = F_3/F_2$.

2. *We assume that $a_n = F_{n+1}/F_n$ and consider the value of a_{n+1}:*

$$\begin{aligned} a_{n+1} &= 1 + \frac{1}{a_n} \\ &= 1 + \frac{1}{\frac{F_{n+1}}{F_n}} \\ &= 1 + \frac{F_n}{F_{n+1}} \\ &= \frac{F_n + F_{n+1}}{F_{n+1}} \\ &= \frac{F_{n+2}}{F_{n+1}}. \end{aligned} \tag{10.54}$$

Since the ratio of successive Fibonacci numbers converges to ϕ (see theorem 10.7), we have a_n converging to ϕ.

The second proof does not use theorem 10.7.

First, we divide equation (10.14) for the golden ratio by ϕ and rearrange the terms to get

$$\phi = 1 + \frac{1}{\phi}. \tag{10.55}$$

Second, from the definition of the iterative procedure (10.47) we conclude that $a_n \geq 1$ for all n (this is also easy to prove by induction).

We also prove the following

Lemma 10.1

$$|a_n - \phi| \leq \frac{|a_{n-1} - \phi|}{\phi}. \tag{10.56}$$

Proof

We consider two cases.

1. *We start from the case $1 \leq a_{n-1} < \phi$. (This case holds, for example, for $n - 1 = 1$.) We subtract equation (10.55) from equation (10.47) to get:*

$$a_n - \phi = 1 + \frac{1}{a_{n-1}} - \left(1 + \frac{1}{\phi}\right). \tag{10.57}$$

The right-hand side simplifies to

$$a_n - \phi = \frac{\phi - a_{n-1}}{\phi a_{n-1}}. \tag{10.58}$$

The right-hand side is positive due to our assumption $a_{n-1} < \phi$ and due to $a_{n-1} > 1 > 0$. Therefore, the left-hand side of this equation is positive as well. Then

$$|a_n - \phi| \leq \frac{|\phi - a_{n-1}|}{\phi a_{n-1}} < \frac{|\phi - a_{n-1}|}{\phi}. \tag{10.59}$$

We see that the difference between the next element a_n of the sequence and ϕ decreases in absolute value by at least a factor of ϕ compared to that for the previous element. We also see that $a_n > \phi$.

2. *The case of $a_n > \phi$ is the subject of an exercise at the end of this chapter.*

Armed with this lemma, a proof of theorem 10.11 becomes straightforward.

Proof

We see that each next element of the sequence gets closer to ϕ in absolute value by at least a factor of ϕ. Therefore, the sequence converges to ϕ.

Note that from these two proofs we can get another proof of the fact that the ratios of successive Fibonacci numbers converge to ϕ. This is an example of how mathematical theorems form an interlocking coherent pattern.

Continued fractions have a number of remarkable properties. To give an example, here is a continued fraction for the base of natural logarithms:

$$e = [2; 1, 2, 1, 1, 4, 1, 1, 6, 1, 1, 8, 1, 1, 10, ...] \tag{10.60}$$

Here the first number is equal to 2. Every third number is given by the arithmetic progression with an increment of 2; other numbers are equal to 1. Not only is this a curious fact, but this representation of e is more economical and converges much faster than the one that defines e (given by equation (6.46)). Table 10.3 illustrates shows convergence for both sequences. We see that the continued fraction relatively quickly approaches $e \approx 2.718281828$, and the sequence $\left(1 + \frac{1}{n}\right)^n$ lags far behind. The continued fraction gets to approximately 2.718286656 for $n = 9$; for comparison, formula $\left(1 + \frac{1}{n}\right)^n$ requires $n =$ 1,000,000 to reach 2.71828047.

TABLE 10.3
A continued fraction compared to the canonical definition of e.

n	Cont. fraction	Approx. value	$\left(1+\frac{1}{n}\right)^n$	Approx. value
1	2	2	2	2
2	3	3	$\frac{9}{4}$	2.25
3	$\frac{8}{3}$	2.6666667	$\frac{64}{27}$	2.370370
4	$\frac{11}{4}$	2.75	$\frac{625}{256}$	2.441406
5	$\frac{19}{7}$	2.58117479	$\frac{7776}{3125}$	2.48832
6	$\frac{87}{32}$	2.71875	$\frac{117649}{46656}$	2.52216
7	$\frac{106}{39}$	2.71794872	$\frac{2097152}{823543}$	2.5464997
8	$\frac{193}{71}$	2.718309859	$\frac{43046721}{16777216}$	2.5657845
9	$\frac{1650}{607}$	2.718286656	$\frac{1000000000}{387420489}$	2.581175

Look at the truncated continued fractions for ϕ or e. Since they combine integers in fractions and additions, they always produce rational numbers (see, for example, the second column in table 10.3). Therefore, a continued fraction may serve as a way to produce rational approximations for an irrational number.

Consider a general expression (10.50) for a continued fraction. It is intuitively clear that the accuracy of such an approximation of P is better if values of b_n for $n \geq 2$ are large. Indeed, values of b_n are in denominators, and having large values in each denominator should make the neglected terms smaller. In that sense, the continued fraction for the golden ratio ϕ has the slowest possible convergence of all because $b_n = 1$ for all n, making ϕ a "very irrational number."

In 1850, the French mathematician Joseph Liouville constructed a number that bears his name and that has the opposite property: its approximations by rational numbers are extremely accurate. The way to construct the Liouville number is as follows. We consider the sequence of $k!$—that is, $1, 2, 6, 24, 120, \ldots$. Then, we form a decimal fraction using the following recipe:

1. Digits placed at $k!$ position have the value of one, and
2. All other digits are zeroes.

Here are some first digits of the Liouville number:

$$L = 0.110001000000000000000001000... \tag{10.61}$$

Here ones are placed at digits 1, 2, 6, and 24, corresponding to the values in the sequence of factorials, and this pattern continues indefinitely. Rational approximations of the Liouville number are excellent. As expected, the continued fraction for the Liouville number contains occasional extremely large values, which help the convergence:

$$\begin{aligned} L = [&0; 9, 11, 99, 1, 10, 9, 999999999999, 1, 8, 10, 1, \\ &99, 11, 9, \underbrace{999\ldots 99}_{72 \text{ nines}}, ...] \end{aligned} \tag{10.62}$$

Joseph Liouville made important contributions to several areas of mathematics and theoretical physics. He also was the first person to recognize the genius of Évariste Galois (whom we met on p. 122).

10.10 Linking Geometric and Fibonacci Sequences

Let's consider the geometric series (9.89), where q is expressed through some other variable x as follows:

$$q = x + x^2. \tag{10.63}$$

Then the sum of the geometric series is given by theorem 9.18:

$$\begin{aligned} \frac{1}{1-(x+x^2)} =& 1 + (x+x^2) + (x+x^2)^2 \\ &+ (x+x^2)^3 + (x+x^2)^4 + (x+x^2)^5 + \dots \end{aligned} \tag{10.64}$$

Let's expand the binomials and see what happens. This is a somewhat cumbersome computation, but it produces an interesting result.

$$\begin{aligned} \frac{1}{1-(x+x^2)} =& 1 + x + x^2 \\ &+ x^2 + 2x^3 + x^4 \\ &+ x^3 + 3x^4 + 3x^5 + x^6 \\ &+ x^4 + 4x^5 + 6x^6 + 4x^7 + x^8 \\ &+ x^5 + 5x^6 + 10x^7 + 10x^8 + 5x^9 + x^{10} + \dots \end{aligned} \tag{10.65}$$

We collect the terms to obtain:

$$\begin{aligned} \frac{1}{1-x-x^2} =& 1 + x + 2x^2 + 3x^3 + 5x^4 + 8x^5 \\ &+ 12x^6 + 14x^7 + 11x^8 + 5x^9 + x^{10} + \dots \end{aligned} \tag{10.66}$$

Explicitly written terms of this sum misses some terms with x^6 and higher powers. Indeed, we only expanded binomials up to $(x+x^2)^5$. This means that coefficients shown in (10.66) for x^6 through x^{10} are incorrect and would be updated as we expand ever-higher terms in equation (10.64). However, the coefficients for the terms from x to x^5 are correct because the binomials from $(x+x^2)^6$ and up in equation (10.64) will not contribute to these terms.

If we limit our observation to the terms that are finalized in equation (10.66), we will see that the expansion results in a power series of x with the coefficients given by the Fibonacci numbers! A proof of this amazing fact is not presented in this book.

10.11 Two Related Sequences

In the world of mathematics, Fibonacci numbers have cousins: Lucas and Pell numbers. Here, we look at the properties of these numbers. Some of these properties will be presented without proof, and proofs of others will be the subject of exercises at the end of this chapter.

10.11.1 Lucas Numbers

Lucas numbers are formed similarly to Fibonacci numbers, but use different starting values:

$$\begin{aligned} L_1 &= 2, \\ L_2 &= 1, \\ L_n &= L_{n-1} + L_{n-2}. \end{aligned} \tag{10.67}$$

The first few elements of the sequence are $2, 1, 3, 4, 7, 11, 18, 29, 47, 76$. Since the recursive scheme for Lucas and Fibonacci numbers is the same, we may assume that there is a connection between these two families of numbers. Indeed, there are identities that explicitly link these two sequences:

$$\begin{aligned} L_n &= F_{n+1} + F_{n-1}, \\ F_n &= \frac{L_{n+1} + L_{n-1}}{5}. \end{aligned} \tag{10.68}$$

Moreover, some properties of Lucas numbers mirror those of Fibonacci numbers. For example, Cassini's identity has a counterpart for Lucas numbers:

$$L_n^2 - L_{n-1}L_{n+1} = (-1)^n \cdot 5. \tag{10.69}$$

A formula for sum of Lucas numbers looks exactly like equation (10.43) for the sum of Fibonacci numbers:

$$\sum_{n=1}^{N} L_n = L_{N+2} - 1. \tag{10.70}$$

Finally, the analogue of Binet's formula for Lucas numbers is as follows:

$$L_n = \phi^n + \psi^n. \tag{10.71}$$

10.11.2 Pell Numbers

Pell numbers are defined by $P_0 = 0, P_1 = 1$ and the recursive equation

$$P_n = 2P_{n-1} + P_{n-2}. \tag{10.72}$$

The first few elements of the sequence are $0, 1, 2, 5, 12, 29, 70, 169, 408$. They can be computed by the closed-form formula

$$P_n = \frac{(1+\sqrt{2})^n - (1-\sqrt{2})^n}{2\sqrt{2}}, \tag{10.73}$$

which is similar to Binet's formula for Fibonacci numbers. The ratios of successive Pell numbers P_n/P_{n-1} converges to the so-called *silver ratio*.

$$\lim_{n\to\infty} \frac{P_n}{P_{n-1}} = 1+\sqrt{2}. \tag{10.74}$$

If $P_n/P_{n-1} \approx 1+\sqrt{2}$, then

$$\frac{P_{n-1}+P_n}{P_n} \approx \sqrt{2}. \tag{10.75}$$

Equation (10.75) is a way to approximate the square root of 2 by a rational number. For example, compare $\sqrt{2} \approx 1.414213562373095$ with

$$\frac{P_2+P_3}{P_3} = 1.4 \tag{10.76}$$

or

$$\frac{P_8+P_7}{P_8} = \frac{577}{408} \approx 1.4142156862745. \tag{10.77}$$

Both these approximations for the length of a diagonal of a unit square have been used since antiquity (without referring to them as Pell numbers, of course).

Just like the golden ratio, the silver ratio $1+\sqrt{2}$ is a Pisot–Vijayaraghavan number—that is, its successive powers become close to natural numbers. For example,

$$(1+\sqrt{2})^{30} \approx 304278004997.99999999999671353. \tag{10.78}$$

But the silver ratio is also a number in the form of $m+n\sqrt{k}$ that we considered in theorem 2.18 on p. 41. Therefore, integer powers of this number will have the form:

$$(1+\sqrt{2})^N = l + j\sqrt{2}, \tag{10.79}$$

where l, j are integers. In the above example,

$$(1+\sqrt{2})^{30} = 152139002499 + 107578520350\sqrt{2}. \tag{10.80}$$

The integer numbers in right-hand side possess an interesting property for any large power N. From equation (10.79) we get[2]

$$(1-\sqrt{2})^N = l - j\sqrt{2}, \tag{10.81}$$

that is, changing the sign in the left-hand side of equation (10.79) just changes the sign in the right-hand side, without affecting the values of l and j. Since $|1-\sqrt{2}| < 1$, raising it to a large power in equation (10.81) produces a very

[2]A proof of this fact is the subject of exercise 9 at the end of chapter 5.

small value. Therefore, $l - j\sqrt{2}$ must be very small, which means that l/j must be a good approximation for $\sqrt{2}$. Indeed, from our example (10.80) for $N = 30$ we get:

$$\frac{152139002499}{107578520350} - \sqrt{2} \approx 3 \cdot 10^{-23}. \tag{10.82}$$

Therefore, the two terms in the right-hand side of equation (10.79) have nearly equal values! In this example, they differ by less than 10^{-11}.

Pell number show some parallels with Fibonacci numbers, just like Lucas numbers do. For example, the following equation resembles Cassini's identity:

$$P_{n+1}P_{n-1} - P_n^2 = (-1)^n. \tag{10.83}$$

In addition, Pell numbers have some identities of their own. For example,

$$\sum_{n=0}^{4N+1} P_n = (P_{2n} + P_{2n+1})^2 \tag{10.84}$$

is always a complete square.

Exercises

1. A generalized Fibonacci sequence starts with $\Phi_1 = a$ and $\Phi_2 = b$ and is defined by $\Phi_{n+1} = \Phi_n + \Phi_{n-1}$. Prove that $\Phi_{n+1} = F_n a + F_{n-1} b$. Prove that the ratio of the consecutive numbers in a generalized Fibonacci sequence approaches ϕ.
2. Consider a similar generalization for Pell numbers: $\Psi_1 = a, \Psi_2 = b$ and $\Psi_{n+1} = 2\Psi_n + \Psi_{n-1}$. What condition must a and b satisfy in order for all Ψ_n to be odd?
3. Prove that $F_{n+m} = F_{n-1}F_m + F_n F_{m+1}$. (Hint: use induction over m.)
4. Prove by induction equation (10.71) for Lucas numbers. (Hint: take $L_n = \phi^n + \psi^n$ and multiply it by $1 = \phi + \psi$.)
5. Prove that $L_n = F_{n-1} + F_{n+1}$.
6. Prove that the number of Fibonacci numbers between n and $2n$ is either 1 or 2.
7. Observe a pattern of even and odd Fibonacci numbers as a function of n and prove that this pattern holds for all n.
8. Application of Binet's formula for F_{n-1} produces

$$F_{n-1} = \frac{1}{\sqrt{5}} \left(\phi^{n-1} - \psi^{n-1}\right). \tag{10.85}$$

Prove this identity by using $F_{n-1} = F_{n+1} - F_n$ and substituting here Binet's formula expressions for F_{n+1} and F_n.

9. Prove that $\lim_{n\to\infty}(F_n/F_{n-1} - F_{n-1}/F_n) = 1$.
10. Prove by induction that

$$\sum_{k=1}^{n} F_k^2 = F_n F_{n+1}. \tag{10.86}$$

11. Use Binet's formula to prove that there exists such N that for all $n > N$ it is the case that $F_n > 1.6^n/\sqrt{5}$.
12. Show that Cassini's identity follows from Catalan's identity

$$F_n^2 - F_{n-r}F_{n+r} = (-1)^{n-r}F_r^2. \tag{10.87}$$

13. Complete the proof of theorem 10.2—that is, prove that $b/(a-b) = \phi$.
14. Complete the proof of lemma on p. 259—that is, consider the case $a_n > \phi$.
15. Use Binet's formula to prove that

$$F_{n+1} + F_{n-1} = \phi^n + \psi^n. \tag{10.88}$$

(Hint: substitute values of $F_{n-1} + F_{n+1}$ from Binet's formula and then factor out ϕ^n from the terms containing ϕ, and ψ^n from the terms containing ψ.)

16. Prove that

$$F_{2n} = F_n F_{n-1} + F_n F_{n+1}. \tag{10.89}$$

using two methods:

(a) Using the result from exercise 3, and
(b) Using equation (10.88) and Binet's formula.

17. Prove that expression $(P_{n-1} + P_n)/P_n$, where P_n are Pell numbers, approximates $\sqrt{2}$ for large n (equation (10.75)).
18. Prove equation (10.69) (the analogue of Cassini's identity for Lucas numbers).
19. Prove equation (10.83) (the analogue of Cassini's identity for Pell numbers).
20. Prove equation (10.70).
21. Use the explicit formula for Pell numbers (10.73) and the discussion of equation (10.81) to prove that P_{2n}/P_n is an integer for any n.
22. Prove equation (10.74) (assume that the limit exists).

23. Prove equation (10.75) from equation (10.74).
24. Prove that the silver ratio $1 + \sqrt{2}$ is a Pisot–Vijayaraghavan number—that is, its successive powers approach natural numbers.
25. We rewrite the definition of Fibonacci numbers (10.1) as

$$F_{n-2} = F_n - F_{n-1}. \tag{10.90}$$

This way, we can start from any pair of Fibonacci numbers, go backwards in the sequence, and define Fibonacci numbers for negative n. Prove that

$$F_{-n} = (-1)^{n+1} F_n. \tag{10.91}$$

26. Consider sequence

$$c_n = \sin \frac{\pi \ln n}{\ln \phi}, \tag{10.92}$$

where ϕ is the golden ratio constant. Since this sequence is bounded, it contains one or more convergent subsequence (see theorem 9.13). Construct a subsequence that converges to zero.

Conclusion

We went through more than 150 proofs. They illustrate how mathematics is constructed. If math is compared to a building, then the foundation of mathematics is a set of definitions and axioms. This foundation bears an intricate structure of theorems that comprise the frame of the building. This frame is rigid and has many levels, as proofs of most theorems use some other theorems.

We are permitted to walk anywhere inside that building, climb stairs and open doors. Some of the doors are old and difficult to open, but none display "No trespassing" signs. We do not have a plan of the building, and sometimes discover a secret passage between two seemingly distant rooms. Some of the rooms have windows to the outside world, that offer different views, from an industrial landscape to the stars in the night sky. These windows are the applications of math to other areas of knowledge.

A great part of such a journey is that we can discover entirely new living areas in the building of mathematics, where no one has stepped in before. This creates the feeling that the building is infinite. The only requirement for walking into a new room is that we must make sure that it is securely bolted to the frame using solid proofs. Then the new room would not collapse and would not take the rest of the building with it.

In this book, we explored but a tiny area of mathematics. I hope it gives you an appreciation of the rigor and beauty that are the unique traits of mathematics.

Further Reading

To explore the subject of this book further, you may want to check out the following publications:

1. *The Nuts and Bolts of Proofs* by Antonella Cupillari [1] is a gem.
2. *A Transition to Proof: An Introduction to Advanced Mathematics* by Neil R. Nicholson [4] goes through methods and types of proofs and illustrates them by exploring sets, functions, and topology.
3. The book by Joel David Hamkins, *Proof and the Art of Mathematics: Examples and Extensions* [3] presents proofs in number theory, graph theory, order theory, and real analysis.
4. Suely Oliveira and David Stewart's book *Building Proofs: A Practical Guide* [5] presents proofs in different areas of mathematics: sets, calculus, Fibonacci numbers, convex functions, prime numbers, and graphs.
5. *An Introduction to Mathematical Reasoning: Numbers, Sets and Functions* by Peter J. Eccles [2] focuses on sets, numbers and arithmetic.
6. Chapter 1 in Kenneth Rosen's book *Discrete Mathematics and Its Applications* [6] gives a thorough introduction on the rules of logic and on methods and types of proofs in mathematics.

Bibliography

[1] A. Cupillari. *The Nuts and Bolts of Proofs: An Introduction to Mathematical Proofs.* Academic Press, 2023.

[2] P. J. Eccles. *An Introduction to Mathematical Reasoning: Numbers, Sets and Functions.* Cambridge University Press, 1998.

[3] J. D. Hamkins. *Proof and the Art of Mathematics: Examples and Extensions.* The MIT Press, 2021.

[4] N. R. Nicholson. *A Transition to Proof: An Introduction to Advanced Mathematics.* Chapman and Hall/CRC, 2019.

[5] S. Oliveira and D. Stewart. *Building Proofs: A Practical Guide.* WSPC, 2020.

[6] K. Rosen. *Discrete Mathematics and Its Applications.* McGraw Hill, 2018.

Index

Only major or first occurrences of each term are included in the index. Theorems are included only if they have proper names.

Printed in the United States
by Baker & Taylor Publisher Services

Printed in the United States
by Baker & Taylor Publisher Services